POWER,
AUTONOMY,
UTOPIA

New Approaches toward Complex Systems

Contributors

Stafford Beer
Ernst von Glasersfeld
Vladimir A. Lefebvre
Dennis L. Meadows
Helga Nowotny
Robert Rosen
Robert Trappl
Len R. Troncale
Stuart A. Umpleby
Francisco Varela

POWER,
AUTONOMY,
UTOPIA

New Approaches toward Complex Systems

Edited by
ROBERT TRAPPL

University of Vienna
Vienna, Austria

PLENUM PRESS • NEW YORK AND LONDON

Library of Congress Cataloging in Publication Data

Power, autonomy, utopia.

Includes bibliographies and indexes.
1. Cybernetics—Congresses. 2. System theory—Congresses. 3. Teleology—Congresses. I. Trappl, Robert. II. Österreichische Studiengesellschaft für Kybernetik.
Q300.P69 1986   001.53   86-12384
ISBN-13: 978-1-4612-9305-7   e-ISBN-13: 978-1-4613-2225-2
DOI: 10.1007/978-1-4613-2225-2

© 1986 Plenum Press, New York
Softcover reprint of the hardcover 1st edition 1986
A Division of Plenum Publishing Corporation
233 Spring Street, New York, N.Y. 10013

PREFACE

The "world" is becoming more and more intractable. We have learned to discern "systems" in it, we have developed a highly sophisticated mathematical apparatus to "model" them, large computer simulation programs handle thousands of equations with zillions of parameters. But how adequate are these efforts?

Part One of this volume is a discussion containing some proposals for eliminating the constraints we encounter when approaching complex systems with our models: Is it possible, at all, to design a political or economic system without considering killing, torture, and oppression? Can we adequately model the present state of affairs while ignoring their often symbolic and paradoxical nature? Is it possible to explain teleological concepts such as "means" and "ends" in terms of basically 17th century Newtonian mechanics? Can we really make appropriate use of the vast amount of systems concepts without exploring their relations, without developing a "system of systems concepts"? And why do more than 95% of all system modelling efforts end in just a heap of printed paper, and nothing else?

Leading scientists from different disciplines, who have different viewpoints and use very different styles in presenting their message were invited to present their approaches to these and to other problems of equal importance: Either as Plenary Lectures at the Seventh European Meeting on Cybernetics and Systems Research at the University of Vienna, Austria, (Professors Stafford Beer, Helga Nowotny, and Robert Rosen (Ross Ashby Memorial Lecture)) or as Invited Lectures to the Austrian Society for Cybernetic Studies, Vienna, (Professors Dennis H.Meadows, Lenard R. Troncale), where my lecture too was presented.

In a time when the possible alternative to dialog across borders is global destruction, Dr.Vadim Sadovsky of the Systems Institute of the USSR Academy of Sciences in Moscow and Professor Stuart A.Umpleby of The George Washington University, Washington, D.C. are organizing a series of meetings of Soviet and American cyberneticians and systems researchers to compare and thus clarify the conceptual structures in cybernetics and general systems theory. **Part Two** is an edited transcript of the lectures and discussions of the panel "Guiding Questions and Conceptual Structures in Cybernetics and General Systems Theory: Comparative Studies" of this Seventh Meeting, by Professors Ernst von Glasersfeld, Francisco Varela, Vladimir A.Lefebvre, and Stuart A.Umpleby, who also moderated this panel.

All scientists innocently accepted the invitation just to lecture and then were confronted with a transcript of their presentation and were also asked to transform it into a publishable paper. All, with the exception of two, undertook the laborious effort to edit and sometimes even rewrite their papers. To them I am most grateful. Probably they liked the idea of this book.

This volume would not have come into being without the help of many others: First, the Austrian Society for Cybernetic Studies, which organized the Meeting and generously supported the preparation of this volume. Then, in chronological order, Othmar Eichinger, who professionally recorded several lectures and many of the often unstructured discussions; Pia Hotko, Gerda Helscher, and especially Karin Schmid, who cheerfully and patiently keyed in the transcripts, then the edited versions, and still were forced to include my editorial changes; Christa Zeller, M.A., who extremely carefully proofread the final version and, in addition, made many proposals for text improvements; Dr. Werner Horn, who was extremely helpful in shuffling around the different variants of the text files and then spent many hours in making them into a camera-ready printout; Dr. Robert H.Andrews of Plenum Publishing, who was a really fair partner during all negotiations; and Professor Helga Nowotny, who proposed the title of this book.

You, the reader, probably had no chance to participate in the very lively discussions. I have thought of a substitute: At the end of the book you will find the addresses of all authors plus short biographies. If you want to get in touch with them, just drop them a line. I am sure they will appreciate hearing your comments and opinions.

I hope this book will give you hours, or at least moments, of thinking and pleasure.

January 1986   Robert Trappl

CONTENTS

PART ONE

Recursions of Power   3
    S. Beer

Not Quite Human: Science and Utopia 19
    H. Nowotny

The Physics of Complexity 35
    R. Rosen

Knowing Natural Systems Enables Better Design of Man-Made Systems:
    The Linkage Proposition Model 43
    L.R. Troncale

Guidelines for Influencing Social Policy through Strategic Computer
    Simulation Models 81
    D.L. Meadows

Reducing International Tension and Improving Mutual Understanding
    through Artificial Intelligence: 3 Potential Approaches 97
    R. Trappl

PART TWO

Steps in the Construction of "Others" and "Reality":
    A Study of Self-Regulation 107
    E. von Glasersfeld

Steps to a Cybernetics of Autonomy 117
    F. Varela

Second Order Cybernetics in the Soviet Union and the West 123
    V.A. Lefebvre

Methods for Making Social Organizations Adaptive 133
    S.A. Umpleby

Discussion: Guiding Questions and Conceptual Structures in Cyber-
    netics and General Systems Theory: Comparative Studies 139
    S.A. Umpleby

Biographies of Contributors 149

Name Index 155

Subject Index 159

Biographies of Contributors 149

Name Index 155

Subject Index 159

RECURSIONS OF POWER

Stafford Beer

Mr. Chairman and the new President, Distinguished Guests, Ladies and Gentlemen:

First of all I apologize for disfiguring this beautiful place with my flag (see below for schematic diagram, which was exhibited in large-scale and colour). It is a flag under which I wish to sail a voyage this morning and I invite you to come with me on what could be quite an exciting adventure. You will have to be brave and adopt such a system's notion as I lay before you.

Now I have surveyed the proceedings of the conference which, of course, these days are published in advance (it is something which always amazes me, because the proceedings are supposedly a reflection of what happened at the conference). First of all, I would like to congratulate the authors of so many diverse and deep analytical statements in many dimensions. I think it is very proper for us to investigate so many diverse things in such detail, and I don't want anything I shall go on to

Fig. 1. Schematic framework for discussion of the social cybernetics for the human condition.

contradict that.   After all, the history of scientific development in our civilization and our era displays a reductionist methodology. It has been extremely profitable to us. It has taken us to the moon, and it has given us huge advances in many, many fields: in medicine, as well as in atomic physics, and so forth. But the price of all this scientific advance, I suggest to you, is that we have finished up with an essentially reductionist model of the universe. And the universe for us, for our civilization, now turns out to be just what science can explain.

I think myself, and I suggest to you, that it is actually quite evident that there is another reality than this one. It is a reality in which we shall not be frightened to face evidence of other things than those things that science can explain. For instance, if you wanted an example, the evidence for what is usually called telepathy is overwhelming; but we don't know how to put it into the reductionist model of the universe that we have. And therefore scientists are frightened to talk about it. I very well remember what happened when the first people walked on the moon and personally conducted an unofficial experiment in telepathy because nobody dared to say out loud that this is what was going to happen. I have spoken with those concerned. There are many other examples ranging from genetics to alternative medicine. Then what I am saying is that we must expect to find, before we are much older, a new synthesis in science. It is being pioneered, as usual, by physics. The account of particle physics and the mathematics that goes with it that is now emerging, not to mention the macrocosmic firmament of black holes and their mathematics too, is going to give us a very different kind of universe. It is one that is not going to be quite so reductionist as the old one.

Now what about the systems sciences? In all of this development we have, I think, a very, very special input to make; because after all our spirit is <u>contrary</u> to reductionism. Having said quite enough polite things about the successes of reductionism, let me now say that if we are going to have a new model then it will be a systems-directed model. It will use the prefix 'syn' a lot: synthesis and synergy outstandingly. Well now, if you will accept, at least for the purposes of our voyage of discovery under this flag, that what I have said so far is at least possible (although it may be a little disturbing), then I next want to suggest to you that the clue to this new scientific search that is systems-directed lies in the nature of <u>invariance</u>.

When I said that I could congratulate the authors of the papers in the proceedings for good scientific work of an analytical kind, I was also conscious (and this is a criticism of everybody, not anybody) that progress in science has always rested in the detection of invariants in systems. This is how we come, epistemologically speaking, to discuss what we loosely call the "Laws of Nature". We detect in gravity, we detect in entropy, and so forth, <u>invariant properties of systems</u>. Now I am submitting to you that we have been very slow because of our reductionist methodology to determine invariance in the systemic world we are looking at. And I think that our great new thrust has got to be in this direction. Now what do I mean by invariance? Just to explore it a little: you know very well that I mean if I say: this expression, $E = mc^2$, will hold throughout a particular domain of the universe. Boyle's Law holds, Ohm's Law holds, the laws of gravitation hold, entropy laws hold. These are the invariants. But let me put to you the question that Ross Ashby, one of our very beloved grandfathers of cybernetics, used to pose to illuminate this question about invariance: To what extent is the Rock of Gibraltar a model of the brain? Do you remember him saying that? He says it in one of his books. Now that is a very strange question to most people. But of course the answer is: If you are interested in physical or

4

temporal extensity, and that is your interest and your only interest, then the Rock of Gibraltar will make a very good model of the brain. That is what I mean by invariance. We are not just concerned with the great laws of nature, but with the things to which we can point and say: "This always happens".

Now, I believe a whole mass of discoveries awaits us in this area. The other thing I want to say by way of introduction is that I have noticed that in our treatment of social systems - not only cyberneticians but other kinds of social scientists - we have very largely neglected the very difficult issue of power. People write about political systems from a scientific point of view as if they had never heard of guns and torture and oppression; just as they write about economic systems as if they had never heard of economic repression and exploitation and alienation. Why is this? Is it that we feel we are clouding science with politics if we address these matters? Well, I can't help it, because if these notions of power are endemic to social systems, as clearly they are, then we have to discuss them. And it is just like stumbling over the evidence in physics for paranormal activity, as I mentioned earlier for telepathy, for strange on-goings within the model 'science'; because in social systems we equally have these facts of power staring us in the face and we don't discuss them.

What is power? Elias Canetti, the great social scientist, has a very simple statement. He says that "power is the will to survive". Now, that has an interesting connotation because it implies a notion of identity that we do not often face up to, either. If we are going to have a will to survive, then <u>what</u> is going to have a will to survive? Something, someone, some institution? We have not spent much time discussing what we mean by power perhaps because we have not spent much time on discussing what is identity. And if you are going to talk about survival - and I have spent most of my life discussing the nature of viable systems, which are systems capable of independent survival - then, I am submitting to you now, we have to start thinking about the identity that is to survive. It is the idea of self I am talking about. And we must expect that <u>identity</u> is one of the invariants that I was mentioning earlier. I have called this talk 'Recursions of Power' simply because, believing that we have to include power in our equation, in our understanding of the universe and especially of social systems - then I am suggesting that the identity that underlies the need to survive and to exert that kind of power will be an invariant that we shall find at every level of recursion.

Now I am using 'recursion' in the mathematical sense of one system implying and being included in the next. It brings me straight to what this symbol is that I have called my flag for this address. What do you think it is? Would anybody like to say what this is? A mandala! Well done; thank you. It is obviously a mandala. (And do you know that C.G.Jung wrote that he had studied mandalas for 14 years before he dared say or write a word about them?) So we have some kind of mandala, and it is one I designed myself, but it works on certain very fundamental principles. But, you know, it looks like a lot of other things too. Surely we have rather restricted ourselves with our systems-diagrams, with little boxes and arrows and so forth. I wanted to give you a richer symbol. For instance this mandala here is a very good diagram of an insect colony. It is a very good diagram of what the alchemists were doing. If you disturb a surface of sand by generating a pure, sound tone, you will make a pattern in the 'liquid' which looks very much like that. I showed the diagram to a very famous historian who said: "Ah, you have modelled the perfect design of a Renaissance city". I could give you many more examples. Another one I like very much concerns the famous double helix of DNA which you usually see from the side looking like a spiral. If you generate a computer view of the double helix of DNA from the end, it looks like this mandala. Now, of course, these invariances are literary devices. I am offering you an imaginative leap here, not a scientific demonstration.

What becomes interesting about this kind of diagram if you really work on it, whether you analyze it as if it were a servo-mechanism, or meditate in front of it as if it were a mandala , the sort of thing that comes out of it is this: that the properties of a system that has identity - and that is what it is really a model of - are such things as self-regulation, self-organization, self-awareness, and in general then self-reference. In consideration of the characteristics of life we added self-reproduction (or so we used to say); but since Maturana and Varela I hope we are concentrating more on self-production than on self-reproduction. Well, these are the kinds of 'self'-things, the 'auto'-things, that we are going to find in our development of notions of identity, and therefore in notions of the power that maintains that identity, whether in the indvidual or in the social system. So that's where the recursion idea comes in, and I will demonstrate its application in a minute. Meanwhile, I hope you all know the work of Maturana that I was referring to under the heading of self-production which, to use his term, defines autopoiesis - the business of 'making oneself' perpetually. This is central to my theme today. As to self-awareness: I believe that many branches of science are pointing to the fact that our big new conceptual breakthrough in science (for which I hope perhaps before the end of the century) will be a proper understanding of consciousness. And I think that physics and biology and the social sciences and perhaps aesthetics - why not? - are likely to join hands together with philosophy and the rest of us in trying to understand what that is. Well, whatever it is, I am urging on you now that this is very much our field, a major field for the advance of our subject.

This is a keynote address, and so I feel the urge to lay before you where I think that we can go when we have set aside for the time being all the individual work that you have done and I have done on individual systems and parts of systems. Here is the picture that is emerging: power lies in the issue of self for individuals and for large social systems. Scientific power in discussing that lies in the notion of invariance, where our findings will apply to both. And that is why I am talking about recursions.

Well, that was by way of introduction. I now want to use this diagram as a model of four levels of recursion. Arbitrarily four: there are thousands. Let us start with the individual, You and Me, the Person. This is a model of such a person. Then we should go to a higher level of recursion of the group of people, whether as a community, like a village or a town, or as an institution like a hospital or a firm, a business. That will be our second level of recursion, and our flag symbol is a model of that too. The third level of recursion will be the nation. Nations turn out to be very, very central to the issue of power in our age. It seems a tragedy really, because philosophers and people of good will of all kinds have offered us the idea of a whole planet, of one world, of a people. But we end up with nations who fight each other in a lethal fashion, both economically and with weapons. And it seems that we have a cybernetic problem here of the reduction of variety; there is just too much variety generated by mankind in total, not to entail the subdivision of mankind into various separate identities. So we are stuck with the historical process that has produced nations and nationhood. That is my third level of recursion, of which this is also a model. Then my fourth level shall obviously be of the planet as a whole. I shall briefly tell you how this model applies at all four of those levels of recursion. That

6

is my second task now. I think a huge amount of cybernetic talk could be made about invariances existing between those four levels of recursion. But we only have an hour here to have our opening session, and I am going to concentrate on just one cybernetic aspect of this model. I hope to make its recursive invariance stick.

Let us first of all consider this model for the individual. The individual - who is this fellow, or this girl? Let us start with the naive definition of a person as what is enclosed in an envelope of skin. If you look at the diagram you will find that represented by the inner, heavily drawn, circle. This big strong inner circle is meant to represent that envelope of skin. Now as good systems people we know straight away - do we not? - that the boundaries that we use to define systems are critical, and also arbitrary and conventional. Most ordinary human beings would accept the envelope of skin as a boundary. We know better! The physicist in us knows perfectly well that a particle that was part of the definition of MY boundary only ten minutes ago may now be in Jupiter, because we know that this particle is a probability smeared across the universe. That's physics talk. Social scientists might say: 'Well, the boundary is not the envelope of skin; this is just a subset of a family which is part of someting else and so forth. So, having made those reservations, let us nonetheless take the heavy circle as the envelope of skin.

Now, you will see if you look a radial splurge of lines filling this inner circle, filling it! In the centre is a blob, and then a radial complex of lines. Now I am taking the blob to represent the autonomic nervous system, and the radials of lines to represent the central nervous system as a whole. And those radial lines, you notice, go right up to the edge of the circle. That means to say that the ends of my fingers and toes are innervated, the nervous system gets there, the nervous system is in charge of this whole thing that I call my body. Now that is a very interesting fact. It is an exemplification of the law in cybernetics that I want to remind you of and draw your attention to today. It is called the Conant-Ashby Theorem, and says that 'the regulator of a system must contain a model of what is regulated'. The Theorem is a manifestation of the Law of Requisite Variety. Now you might say: 'Well, that is self-evident'. But the funny thing is, you see, that when we get to our other levels of recursion, when we begin to look at social systems, we very soon find that we do not obey the cybernetic rules. The individual, however, considered as the envelope of skin with a nervous system - do note! - is capable, and does obey the theorem. It is capable of containing the variety generated by the body. So if I fling my arm out there I can still move my fingers. My central nervous system does not say: 'I stop here; you fingers are on your own out there'. The brain, in short, has the model - not only the brain, the whole nervous system has this model in it.

As I said, it is very useful to distinguish between the autonomic nervous system and the rest of the nervous system. The reason why it is so interesting is thir: If you are going to have a very high variety model in your regulator, then much of it must be autonomic, which is to say: it's self-regulating. Otherwise, of course, we would have to put most of our conscious effort into keeping the system going - keeping the heart beating; stopping from falling over; bringing the hand back when you have thrown it out; and so forth. I repeat that when we get to social systems we shall see how much this principle is disobeyed, the principle that we use and exemplify in the body.

Meanwhile, however, let me progress to the next level. I have been talking about the heavy ring and the radial lines that reach its circumference. Look next at the second circle within which all this is em-

7

bedded. This is the part of the individual which is not fully realized. It stands for the capacity to do something. For example: you want to run a marathon - can you run a marathon? - it is 26 miles, you know, it's a long way - I bet, not many people in this room could run 26 miles. But they could if they trained. I can't play golf, but I probably could if I practiced. Now, these simple examples of things that one might do with oneself can be extended to much more serious matters. You could all learn Sanskrit, if you gave yourself the time and the motivation to do that. There are physical things to do, there are mental things to do, there are, indeed, spiritual things to do. You can set yourselves spiritual goals as well as mental goals as well as physical goals: that is the individual defined by the second circle. S/he is no longer just the envelope of skin, but the aspiring individual. The tiny circles on the ring's circumference stand of course for the goals themselves. Now then, where is the Conant-Ashby theorem in this? Please, think about it very hard. Your model of yourself and the regulator that you have for regulating yourself does not include the things I have just mentioned - until you put them in. And if you devise a circle of aspiration for yourself and say: 'I am going to aspire to do this, that and the other', then you will have to change your model of yourself, will you not? If you want to run a marathon race, you are going to have to take up jogging in the mornings. Right now you do not jog in the mornings; so your model of yourself has to change·in order that you become a jogger to yourself.

Now we are already beginning to find some very important lessons out of this analysis. They will serve us in good stead as we move through the 'Recursions of Power'. Remember that I am going to talk about invariance alone this morning. It is invariant in an identity, in a system of self, that the things the system is capable of, but is not yet realizing, are not initially included in the regulatory model. And then people try to do things without changing the regulatory model. Now, how have I depicted that in my diagram? You will see that the lines radiating out from the centre, which went to the edge of the envelope of skin (and therefore provided adequate regulation for the corporeal body) do not quite extend to the edge of the second circle. This is a diagram to represent the fact that we know that we can control ourselves further than our existing way of living, but we are not quite certain how to do it. And if I set about running a marathon, I am not at all sure about how I would control myself, recreate the model of myself, and push those radial lines out to discipline myself (we would say, in the case of a marathon). So we have a control problem of requisite variety as soon as we leave the ostensible self. The ostensible self is the inner circle, the capable self is the expanded individual. Take a further look at the explicit goals, the tiny circles. The capacity to run the marathon becomes a goal, the wish to learn Sanskrit becomes a goal; and we can define those goals, and we can say 'damn it, I'll do it!' The most beautiful book ever written on calculus, I'm just remembering, begins by saying: 'what one fool can do, another can.' So you decide to learn calculus if you can't already do calculus, and that is an explicit goal. And now you will see those black lines, tangential to the ostensible self, which are relating your goals. Notice how powerful this model is becoming. If you want to go from the ostensible individual self to the expanded self, then you will have to increase your regulatory model as exemplified by the radial lines, and you will also have to build - what do we say - 'strategies' for your life, which are the black relations between explicit goals. But you are already in considerable cybernetic difficulty, because of the law of requisite variety. This is why most people, I submit to you, fail. I am talking now of psychology, if you will. People fail in their goals; they join correspondence courses to learn Sanskrit and then don't do it, having spent a lot of money. They buy golf clubs and so on and say: 'I am going to play golf' and then don't do it. Their model of themselves is

8

defective, and the regulatory system is trying to disobey cybernetic laws.

Now I come to the outside ring - the third ring of power. What is that? Well, it's simple. Having discussed the goals we can distinguish and therefore make explicit at the second level, this level says: 'well, there are goals that we can't detect, because we don't know what our ultimate capacity is.' And for this final ring I use Aristotle's word, which I would like to bring back into circulation in science. It has been mostly out of use for 2000 years. Aristotle's word was 'entelechy', which means the fulfillment of promise, of potential - the total fulfillment of potential. The final circle of the diagram is incomplete; that indicates our uncertainty about the boundaries of entelechy. So leaving psychology we come to the area where preachers and gurus, all those kind of people, are saying: 'Look, you, Sir or Madam, have much more potential than you know; do something with yourself, beyond the goals that you can distinguish, and grow to your full self!' Now I am using the word 'ostensible self' for the inner ring, and 'potential self' for the second ring where you can distinguish goals, and 'entelechy' for the final affair. Most people live through their lives without ever contemplating entelechy, as you know. And who shall scorn them for that; most people are starving or rotting in jail. We need to look at the statistics of this planet, as we speak here in comfort and ease. The mass of humanity is in terrible trouble, and they, perhaps, do not have time to contemplate entelechy. We do; and maybe we have a responsibility to think about it on their behalf as well as our own.

Now, before we leave the individual, I want to say that I am a scientist despite a lot of philosophic talk here today. And, of course, whatever we do with ourselves and however mystical we may sound in discussing to what heights the human being can aspire, what the human being does is mediated by a control system. Outstandingly, this is the central nervous system, as I mentioned at the outset, the brain. One of the things I would like to leave with you out of this part of the discussion concerns the way we discuss the brain. The brain is always discussed - have you noticed this? - in terms of the available technology of the day. This is a rather ominous thought. We seem to use the latest technology we have got to talk about the brain because the latest technology looks so new and so good and it appears to be at the forefront of scientific understanding. It is very far from being 'absolute'. Let me remind you. Aristotle thought that the brain was a machine for cooling the blood. Well, it is, you know, with all that surface area. But that's not its primary purpose. Let's move on rapidly. Descartes discussed the brain in terms - do you remember? - of the fountains in the King's gardens. His technology was hydraulic; so the brain was squirting juice all over the place. You get to Locke - and the great age of the advancement of mechanics. Locke talks about the brain in terms of nerve processes having little tiny wires inside them which run over invisible pulleys, and that whole thing is a mechanical artefact. You get to von Neumann, at the time when computers were becoming THE thing, and everybody then used the phrase of the 'electronic brain'. So the brain now became an electrical switchboard, and certainly bits of the thalamus look a bit like that. Warren McCulloch, my beloved mentor, used to say that the brain was a three-pound electrochemical computer running on glucose at 25 watts. That was another way of confronting people with a physical rather than a metaphysical reality, using the kind of technology that was available. Warren was that kind of analyst of the brain.

I consider that our discussions of the brain are going to be crucial to our understanding of selfhood, of consciousness, of identity and of power. My theme is building up, isn't it? I want to leave a special blessing, therefore, for neurocybernetics. This field must advance, and I am conscious that I am speaking in Vienna where the Cybernetic Society is founded in the person of Professor Trappl in a medical school. As to my own role in this - I feel I must mention it - my own mathematical model of the brain was done in the late fifties. Very few people here, I think, will know that model; it was published in 1960. It depended on the following idea that, since causality in the brain is a very difficult thing to follow, what we should try and realize is that the sensory part of the brain and the motor part of the brain - whatever the causal connection between them - must in some sense map onto each other. This mathematical model was set-theoretic: it made no attempt to indicate transfer functions that nobody understands. The interesting thing about that, when we are talking about the technologies that we use to describe things, is that the model generates the notion of the brain as an interference pattern. And this was in the late fifties. I had not then heard of holography. By now, not surprisingly, I cannot think of the brain as anything other than a hologram. I will return to that later.

Well, so much for the individual, given that I am just using this model as descriptive of particular invariance: Now let's pick this up at the next level of recursion and ask about a social system such as a community or an institution. Have we, in fact, got some invariances out of our discussion of the Conant-Ashby theorem? I want to repeat that we could do what I am doing now for the Conant-Ashby theorem for at least another ten basic principles of cybernetics, but we haven't got time. The point is only to demonstrate that there ARE invariances. Now, in a social system, what is our heavy inner ring? The ostensible system, the accepted system, which depends on the definition of functions and boundaries that 'everybody knows', is the answer.

Take a system of travel. You want to run a railway? Everybody knows that the railway has some tracks and has some cars on the tracks and engines and stations and things; so this is what you have when you have a railway. That's the ostensible system. However, I said about the body, well, if you are particle physicists, particles in the envelope of skin will soon turn out to be on Jupiter. Equally it turns out, if you are a management scientist, which I have been for a lot of my life, then the ostensible system of the railway is not as obvious as you would think. When I found myself advising the Canadian Railways, I certainly expected that all the hardware I just listed would constitute their ostensible system. But I soon found out that the one thing they didn't have was any of these things. They had to hire them. Now there is a very great surprise; but anybody who has done managerial cybernetics comes to know that the system that the management think they are operating is not the system they are operating at all. Many of them never discover this.

You take health. 'Everybody knows' that a hospital is a place for curing people. What would you say, systems-ladies and -gentlemen, if you found a hospital whose whole output was a succession of coffins? I tell you what the authorities would say about that situation. I am giving you a caricature to make my point: Here is a hospital and all the people coming out are dead. Hm? What the establishment says is: 'Well, we have had a bad day or a bad week, or a bad month. But this is an imperfection in the system. We shall put it right.' They never ever think of saying: 'My God, we have a machine for killing people!' It is because everyone knows that 'this is a system for making people well'. When I joined the world's biggest publishing company, I asked them what business they were in and they said: 'You fool, we are in printing and publishing.' But if you look at the assets of the company, as I said to them, they are in the business of real estate. They owned 92 of the prime sites in London; all their assets were tied up in buildings. Nothing to do with printing or publishing.

This is the kind of thing that you find out if you do research in cybernetics in the management field in communities and institutions. You discover a lot of difficulty in defining the inner circle of the ostensible self. And, if that is so, you discover even more difficulty in defining the radial lines. What of the radial lines? Where is the regulatory model? If your epistemology is such that fundamentally you do not recognize the system you've got, what is your hope of defining the regulatory system that will control it? Very little. That has maximum bearing in the case of the second recursion of the community or the institution when we come out to the second circle, because that is the self to which the institution is aspiring. And these days it has a whole technical apparatus to get it there, called planning. New technology, planning, Year 2000....all of this, waiting to take us out to the goals we are setting out to reach. The trouble is, of course, the Conant-Ashby theorem. If we don't have the model of the system we are now regulating right, how much less do we have the model of the radial lines that are reaching out to the second circle. It follows that most of our planning is directed to building a system that could not possibly work if we had it. I speak from a lot of very tortured experience in this regard.

But I must press on. What about the entelechy in the case of the community or the institution? Those of you who have studied the field of planning, I am sure, will know Ackoff, who is one of the doyens of Western planning, and his theory of idealizations. Now, he says, if you were in a university and wanted to know the future of the university, the entelechy of the university, then, by all means, don't start from here and say: 'We will improve this building, we will add courses, we will push outwards'. That is part of the aspiration level of activity. The entelechy is concerned with saying: 'Just a minute. This university is a historical accident. What do we want of a university in the year 2000?' This is the idealization that you design and then you say: 'Well, that's what we want. This is what we have got. How do we get there?' Notice that in using this approach the regulatory system gets to be designed intrinsically with the institution itself, and is not therefore the necessary victim of Conant-Ashby. Those are some of the considerations that apply to the second level of recursion.

Let's take a quick look at the next one, the nation, as I mentioned it. You should be getting familiar with this method of arguing by now. The ostensible nation is the historical nation that we have, the accepted and recognizable national ethos. Now, I have worked in about seventeen countries, and the first thing everybody tells you is: 'Our country is quite different from any other country. We are like this - and proud of it.' Well, I am used to that, because any company will tell you the same thing. They say: 'Don't come here with a lot of business theories, our company is unique.' And of course these dear people, they are all unique. But it does not alter the fact that there are a lot of invariances, such as you go out of business if you don't make a profit. In the national case, if you don't obey the capitalistic rules of the IMF, you don't get the next loan to pay your interest on the last loan. So at the national level the first question is: does the Conant-Ashby theorem apply within the inner circle? Do we have a model of what the nation is included in the regulatory apparatus?

Can you not by now see how this arguing goes? The law is a product of history - is perpetually out of date - is perpetually incapable of providing the regulatory model. So legislators spend all their time propping up the law, passing new amendments to the law. The finance act in most countries is a great big mess of amendments. We try to disobey basic cybernetic principles even at the level of ostensible selfhood. When we get to the second circle in the nation, we find the nation talking about its goals - and it is doing that, of course, all the time, because that's what national politics is about. This is how presidents and prime ministers get themselves elected, they say: 'We are going to do this!'; they have no hope of doing what they are saying, you know that, the ordinary citizen knows that. We cyberneticians, for goodness sake, have the precise reasons why they cannot do what they are saying. It has little to do with the cut and thrust of political debate as displayed endlessly by the media. It is because they don't have requisite variety; they don't have the regulatory model to do it, still less the regulatory machinery.

I mentioned to you the importance of distinguishing between the autonomic and the volitional parts of a nervous system. You think about that in the nation. The constant tendency of people interested in power, which has to do with their own and their party's self-survival as against the national good, causes them to rob the system of autonomy systematically and to centralize. We have got this going on in my country, in Britain, right now in the most preposterous fashion, whereby local autonomy is being lost and the whole nation is getting run from the middle. This is very - if you want political words, you start talking about fascism and things like this - it's very uncomfortable, but if you want to stick with cybernetics, you say: 'This cannot work, because ...!' But unfortunately none is likely to listen in Britain. I spent nearly all last year in Mexico, and the invariances were very apparent - for different reasons, of course, because the national ethos is different. I made a systems analysis of the current president's political intentions, and I isolated seven major objectives of his presidency as revealed in his speeches and his book. Then I made a systemic model of that, and I found out just what cybernetic principles needed to be applied in order to achieve these ambitions. They simply do not have those things in place, and they must fail. I do not care how newspaper people presently conceive this. It is possible to demonstrate cybernetically that the current ambitions of Mexico will not work - and this is before you get to corruption.

Now, you know, corruption is a problem in many of our countries, and I just want to say this about it, that I regard corruption as a <u>systemic</u> failure. I do not believe that men and women are worse in one country than another, I mean in a moral sense. In India, any good cabinet minister will tell you that the Indian people are corrupt as individuals. And if you ask them 'How?', they will say things like: 'If you filled a train with grain in Bombay and sent it to Delhi, I guarantee, it would end up empty at the other end. We are morally corrupt people.' To which my answer very strongly was: 'Don't be ridiculous, this is a system! You have got lots of starving families beside the railway lines, and they will take the stuff off the train; that is wholly predictable, that is a system in operation.' And to quote to you now one of the aphorisms that I always use and hope to make famous: 'The purpose of the system is what it <u>does</u>.' So don't ever let anybody tell you that the purpose is something other than what you see. If a hospital is producing dead people, remember, then it is a machine for killing people. And if everybody takes the grain off a train, then that is because there are starving people. In Mexico, if you have massive corruption, it is because the system dictates that is what there should be. The reason for that in Mexico is very evident, it is right before your eyes. There has been a permanent revolution for seventy years, and a party which actually calls itself the 'institutionalized revolution'. It can remain in power only - and this is a systems point - by fixing the elections. So you are going to talk about moral corruption? Especially when the president has a personal campaign for moral renewal.

This is how national systems - I am giving you only a   sketch   - get
themselves into   such   a   mess, because their 'tiny circle' goals conflict
with the actual system of   regulation   in   the   model   of   the   regulatory
process, which   is   itself   embedded   in   the constitution and in the law.
Constitutional laws are powerful, and you have to change them, not try and
go around them.   I could talk at very great length about that, but I   must
not.   We   need   here obviously in the 'aspirational' circle a new model of
progress.   In developing nations, it is   extremely   important   that   those
definitions of   progress   are   made   by   the   nation itself and not by the
people who want to exploit that nation, which is what has been   happening.
The paradigm for progress and the regulatory system to go with it has been
prescribed by   the   very people who want the raw-materials and the markets
of that nation.   And they are the very last people who should   be   allowed
by that nation to make those specifications.   At least, that is my suggestion to   you.   As to the entelechy of the nation, well, I was already coming to that in what I just said: The thing is that the entelechy now does
not stand for idealizations as it does with the   institution;   it   stands
for the   loose   collection   of things that we call utopias: better societies.   And they and the final circle, you see, are   not   defined   at   all.
You can   have   a national goal at the aspirational level to build a dam or
change the educational system, but as to the   entelechy   you   don't   know.
And so my appeal here to the Third World people is: 'Please, don't import
a whole lot of philosophic rubbish along with the plastic rubbish that you
find yourselves importing from the rich world.'   I watch with great despair
the wonderful   cultures   -   to   take the two nations I have mentioned - of
India and Mexico, dissipating in the face of the   importation   of   plastic
and computers and refineries and things of this kind.

I turn now very quickly to the planetary   level   of   recursion.   And
here I   will make a very fast mapping indeed of this model - from Teilhard
de Chardin.   Do you know him?   - many of you do, I'm sure: 'The   phenomenon of man'.   His model will map straight onto this model.   The ostensible
controlled self of this planet is the geosphere, as he called it;   that is
to say,   the   ball   of rock with a molten interior that we call the Earth.
And, of   course, that does have a proper regulatory system: the   gravity,
the waves,   the wind, the way water and air behave - all of that is a very
firm regulatory system.   It has its own model, and obeys the   Conant-Ashby
theorem.   The   next   level, the second circle, is what Teilhard de Chardin
would call the biosphere, which is the green   envelope   of   living   matter
that covers   the   geosphere.   And   that   itself , of course, has the most
wonderful regulatory mechanisms - in homeostasis, and all of   those   kinds
of things   which support that, and have supported it for millions of years
until we came along and perverted it.   And I don't have to preach to   this
kind of   audience about what we are doing to the biosphere through lack of
understanding of the regulatory models.   We create the dustbowls,   we   use
too much   DDT;   and, above all, we use too much napalm and too many bombs.
We are disrupting the beautiful regulatory mechanism that nature has.   May
I remind you of the Gea-hypothesis, which says: 'The world is actually   a
big living world'.   We are breaking that up through lack of application
of the knowledge we people actually have.   Think of the responsibility, my
friends!   And as to the entelechy: Teilhard talked about the noosphere   -
from the   Greek   'noos'   -   mind   - where he envisaged my final circle, an
entelechy of an expanded consciousness, of perhaps a   world-consciousness,
rather like   the Jungian universal consciousness, only much more than that
in Teilhard's case because of his very great spiritual overtones.

So now: I have shown you how we can use this model at the individual
level and the communal level and the   national   level   and   the   planetary
level, and   how   we   can   perceive invariances.   I am not writing the book
about all this in an hour.   I am showing you only that this   is   possible,
and the   sort   of   thinking   needs to do it.   Now, I want to end by making

good use of this. I see that nobody is leaving me to lunch - so perhaps you will bear with me for a few more minutes. Because having set up this multiple model, ladies and gentlemen, I want to talk to you about the nature of change - in a very unusual fashion, because I am using my mandala here as my guide.

I have been a student of Eastern philosophy for close to 40 years. I am beginning to say out loud now what I have long suspected: that the Eastern philosophies have been based on the notion of system for 5000 years at the very least, and we people are only just discovering it. So let us be a little respectful of that. I want to end with a different analysis of my four levels, and it begins with the notion of change. Now in the West, how do we think of change? First of all, our change is time-dependent. We have a theory of causality which, as a matter of fact, was blown to shreds by David Hume hundreds of years ago, but nobody seems to have noticed, for we still have a causal model of change. So what do we do? We make an analysis of the facts. Think! We make a personal inventory, we say: 'I am like this; I wish to change, I wish to go to the 'tiny circle' goals, and to run the marathon; or I wish to expand my consciousness'. In a community, we say: 'We must change this, that, and the other.' You see - how it fits. So we find out the facts, and then we prescribe our intentions and we say: 'Well, we will make this change; it's going to cost a lot of money. So we must make a budget.' And then we find that the other things that are happening in society mean - like having to have bigger bombs - it turns out that we suspend that budget. We don't actually make the change at our personal human level either. We put it off; we are too busy, and the family makes demands on us; so we don't do it. So, the change is time-dependent. It amasses facts, it says: 'We are going over a period of time to be different - and it is going to cost, it's going to cost time and care and attention and probably a lot of money'. And then we don't do it, as I say.

Now, what is the confrontation there? We say one thing and we do another. But the purpose of the system is what it does. So all of this is so much nonsense; most of what we put into our plans, and especially the good intentions for ourselves and for our society - all of this is time-dependent. We never have the time. The Eastern approach, on the other hand, speaks of change quite differently. It is not a time-dependent phenomenon. Change is a way of realizing yourself. It involves immediate and total confrontation of reality. An Eastern thinker would say: 'We don't talk about change, and generating plans, and all of that stuff because - if we center ourselves properly and confront the truth - then the truth is thereby different; ipso facto.' Now, we have heard of Heisenberg. We should know that this makes a lot of sense. The confrontation of what is, changes it.

I want to run through my four examples in this light. I mentioned that I see the brain as a hologram. Now, a hologram, you know, does not obey ordinary spatio-temporal laws. And the very first thing that anybody who has done work in yoga or any other spiritual discipline - Zen, for instance - the very first thing that he knows is that he is outside the spatio-temporal distinction. An experience called 'Satori' in Zen-Buddhism is essentially that. It is a glimpse of the reality that does not have these Newtonian bounds on it. Now you will see why I started by saying that we were confronted with certain things like telepathy for which there was a mass of evidence, but which did not fit our model of the universe. So I am saying that at the individual level you are going to find - if you use this model - a completely different account of yourself and a completely different way of handling yourself. You want to give up smoking; you want to give up drinking? The Western model of change says: You get the facts, you know how much you are spending, you know how much damage you are doing to yourself. You say: 'I must change this!' You don't do it. Most people don't do it, because - why? isn't it perfectly simple if you look at those systems persons - it is because they like it! It's that easy! So the intellectual part of them, saying: 'Let's give this up!' is not the real part, which just continues as before. The Eastern way, for the individual, is to confront the toxicity of the alcohol and the tobacco and just - stop.

If you want the evidence that this thinking is correct, do realize that Alcoholics Anonymous (using a model essentially from Adler - we are in Vienna) perceived that if you say: 'I am going to change, I am going to give up drinking!', you instantly create in yourselves all the resistance required to overcome that good intention. Something in you is fighting it and saying: 'The hell with that! I am not going to do it!' You have a battle, and you lose the battle. Whereas, if you confront the issue in the Eastern fashion, there isn't a battle at all. There is realization. That is quite a different experience, and some of you must have had it. I hope everybody will have had it, but I think, not.

What happens if you apply this Eastern kind of thinking for the individual to the other levels of recursions and the other fields that I have been talking about, the other selves: community; nation; planet. I will give you just one example of each because I really ought to stop fairly soon. Within the community, take the example of penology. What do you do with criminals? Now, we are all citizens; all of you must have some knowledge of what we do. You know, for instance, that the talk about deterrence is fictitious; there is no scientific evidence that the ways we treat criminals deter them - unless you actually execute them (that deters them). So we know that what we do, does not work. We know that it is appallingly expensive. We know other things about it too. So we keep on saying: 'We will reform the penological system!' It's going to cost a lot of money, and then the budget goes down some other drain. Typically in our society, it goes into the industrial-military complex. So criminality persists, and penology persists.

Now, I want you to try the experiment of using the Eastern way of looking at this, which I have been talking about. What would happen, if as a society we managed - not to talk about budgets and resistence, this is a way of not getting change - but to say: 'Let us confront the reality!' What happpens? I don't know anything about Austria. But I have just come back, I was at the end of last year in California, and I know about the penal system there. The first thing that happens to a young man sent to prison in California is that he is raped. And that is quite general. Because we put people in one-sex prisons, then that is a systemic consequence. But it is also a fact. Now, supposing we use the Eastern method and, instead of saying: 'We will change this!' and not changing it, supposing we said: 'Now let us confront our reality!' What would happen, my friends, if a judge said to a young man: 'You have been caught stealing 20 Dollars; I sentence you to be raped!' Because that is the truth if you want to confront the reality. We should have a bit of an outcry, I suspect. We should have California up in front of the court of human rights. It's a very different perception, isn't it?

I have been trying out this way of Eastern analysis of our social problems and getting all sorts of shocks like that, and I wanted to share them with you. Take it at the level of the nation. I was talking about Mexico just now. What happens if we confront the reality? Big government contracts: let it be confronted. Now what do we do? We ask for tenders. Tenders must specify amounts to be spent on materials and machinery and labor and so on, and how much on bribes. If you did that in Mexico right now, you would have a third of the money tendering for bribes. What a

confrontation that would be! And quite a useful one, too. Because it would indicate that the system does not work without being oiled with this particular oil. Maybe we would learn a whole lot from that instead of wailing about the moral consequences. I told you how suspicious I was of the moral argument. The moral argument can be applied only to the individual, ethically. It is not a recursive invariant - at least, I don't think so.

At the international level, by using this approach, we should get new models of self-regulation. Then we would see - and I believe I can see, but I find it very difficult to express in Western terms - why it is that we have got a system which transmits wealth from poor nations to rich nations, although the rich nations keep passing resolutions saying that they want to make it the other way round. They want to make it the other way round, but they do not have the regulatory model to do it. The regulation is in the hands of the International Monetary Fund, and, in general, of banks. And their model is going in the other direction. The last loan that I saw being negotiated while I was with the Mexican government had a cost attached to it. There is the money for the loan, right. Then there is the money for rescheduling the debt, right. There is the money to pay a whole bunch of lawyers, experts, accountants, economists - you name it - publicity people. The net result of that loan was - without coming to the question of interest - that the cost of getting the loan was exactly the cost of the loan. Can you credit this? I mean, we are collectively responsible for this kind of thing; then we blame Mexico, having made them do it. If you went to your own bank, I don't care which nationality you are, and said personally to your bank manager: 'I cannot pay the interest on the loan you have given me, please give me a loan to pay the interest', he would throw you out. And yet all our international affairs are conducted on that basis. Usury is not a strong enough word for all of this. My dedication to the third world comes out of knowing it at first hand.

Well, the entelechy for the nation: I have said a little bit about it already. It has to be self-referential. It is no use taking on some model from somewhere else. Here is a very small example of that which the Canadians here may recognize: There was a very interesting project on the poor Eastern seaboard of Canada, where everybody was sitting on the doorsteps of their houses saying: 'Look at us! We are in despair, nothing can be done'. And a film team went around with video-cameras and filmed everybody and asked them what was the matter. So everybody said: 'Well, look at me; I can't do anything'. They edited this film, and they showed it to the whole community in the village hall. Can you imagine the impact? Now, this is the Eastern approach again. It fits my Eastern model because it involves self-confrontation. If you were sitting in the hall and the film was running, and every single person in the room was sitting there saying: 'I can't do anything', you suddenly realize that, perhaps as a group, you can do something - because you have confronted a reality. That happened in that pilot project. But, you see, the thinking is so different - people don't take off and do these things, as they should.

The example I would like to give you at the planetary level concerns unemployment. Again, if we really confront things, what do we find? There are about 12 billion people going to be on this planet at the turn of the century, about a billion of them unemployed. A billion people! Now, what I have to say to you is that there is literally no way in which jobs can be created for those people. It is just no use pretending that we can do it. Confrontation of reality! We cannot do that! In 1955 we amplified technology, that is to say, automation, by 20 times with labor. By 1970, it was 10 times. Microchips say that a third of the present labor will be required - even with the need to service equipment - by the year 2000. Sixty percent of European youth will never have a permanent job. These are factual extrapolations of the regulatory system we actually have; not the one we would like to think we have. Well, those are the sort of facts that lead me to say that we will have <u>insurrection</u> if we don't get a new <u>perception</u>. We simply can't afford to continue with the concept of employment, however many models we have based on work ethics and all that moralistic stuff. We don't need that degree of employment, and we must stop putting a social stigma on unemployment quickly! Immediate realization! These are the things that we have to work for.

Summing up, we can see the range of needs. In the individual it is for the redefinition of life-style for himself; in the community it is for the redefinition of the purpose of the community, in the nation for the redefinition of progress , and in the planet for the redefinition of such basic things as employment and the inevitability of war. Let me end by saying that I hope that we will try to put these choices back in the models of regulation from which we have taken them out. It is a priority to get back choice. Now I would like you to know that the great Eastern teachers whom I have evoked today say that there is <u>no</u> choice really; that a clear spirit has no debate with itself. But I can give you a scientific explanation even of this dilemma, this apparent conflict between the teaching and the facts. You know, we have game theory. In a game of complete information there is no choice. At the entelechy, we would <u>have</u> complete information. Meantime, we have not got it. We have complete information in the game of chess, so theoretically we should be able to say: 'You are white, I resign'. But we can't do the sums. So here we are, poor people, unable to do the sums and looking again for choice. So my message is: We have to <u>do</u> some of these things - not just think and research about them. We have to try and put this whole big stuff together in action, somehow. Now I don't know how we are going to do that, but I do beg you to think about it.

You know, it is not enough just to be a professor. I have all my life tried to keep half of my activity in the domain of <u>action</u>, and I recommend that course to you. The great teachers I am evoking again said a lot about this. They didn't much like professors. Jesus said: 'By their fruits you shall know them!' The Buddha talked about professors as 'the herdsmen of other men's cows'. Mohammed said that a professor was 'an ass bearing a load of books' (though this is a bit rude). So I want to leave you with the thought: 'We have to do something!' And I hope that one of the products of this conference will be some prescription for action - as well as the collection of theories - in the context of my 'Recursions of Power'.

Thank you very much!

NOT QUITE HUMAN: SCIENCE AND UTOPIA

Helga Nowotny

1. THE NON-EXISTENT SCIENCE OF UTOPICISTS

While preparing for this contribution I went to see a film: Sans Soleil by Chris Marker. In 100 minutes a dense collage of visual poetry is presented to the spectator, accompanied by an equally dense essay of impressions collected in Japan and Africa. Japan has been chosen as one possible society of the future, representing what the film pictured to be one extreme in the art of survival of a civilization yet to come. What fascinated me was the utopian touch that was carefully and yet emphatically, read out of the present: the music of video-games, for instance, as the constant, underlying musical theme of a buzzing metropolis; a description of how these games were programmed and how a new collective language of imageries was in the making, coding memories and thus providing the essence of a future collective unconscious. Interspersed with everyday scenes, celebrating their banality and uniqueness at the same time, the film cautiously proceeded to construct an imagery of a future, in which humankind continues to evolve, guided by the computer and computational thinking. The emphasis was put on the collective mind, and not the individual, in the making, and how this new form of technology-based consciousness would interact, shape and be shaped by what the film-maker sought to single out. Japanese society was predisposed, in his view, to serve as a model for survival, because it knew how to balance high technology with the mechanism essential for survival - social ritual. Whether these involved prayers for animals or for the spirit of material things, ceremonies of purification, of expressing joy or channelling aggression, mind and - the social - body were pictured as meeting in a gracious, convincing and yet for a Westerner deeply irritating way. The film made no concession to the Western image of Japan; no allusion to the race for the fifth generation of intelligent computers or to the hot economic climate of intense competition appeared, nor any of the themes that figure prominently in the current Western debate - control of data banks and fear of more comprehensive surveillance through state or large corporations; intellectual property rights and the issue of secrecy; the possible isolating effects of the new technologies when substituting standarized expert systems, the artificial experts, for conversations with human experts or with friends. And yet, in its non-intentionality, its obvious digression from the dichotomous mode in which future developments are often presented in the West, the film offered a much more convincing image of what one possible future in the mind- and computer age might look like, than any other account I have come across.

Societal imageries of possible futures are not a thing of the past, although the grand visions of entire societies to be built have apparently given way to a much more fragmented view, either built around minority groups in society or split into a myriad of individualized micro-utopias (for more details, see Mendelsohn and Nowotny, 1984). In running briefly through the history of utopian thought, I would like to focus on the inherent tension between science and utopia, as well as on some of their commonalities. Next, I will turn to the field of AI and robotics, as an interesting example of what actual developments in this rapidly evolving field of research and applications mean for the utopian-dystopian scheme of projecting future developments. Finally, the question of the tension between science and utopia will be taken up again and I will ask whether we are not witnessing the emergence of a new utopia - one that can tell us, perhaps, more about the present than about the future, but which will not fail to influence what the future will look like.

The still fermenting field of AI, cybernetics, systems theory and their application - as also the occasion for this scientific gathering demonstrates (Trappl, 1984) - has perhaps as no other recent field of growth of scientific knowledge and engineering, both, such a long history in utopian and dystopian thought and actual developments that appear to have the potential of realizing what has been anticipated in a negative and positive version. It offers an experimental work-shop in utopian-dystopian thinking, inviting comparisions, for instance, of the old and recurrent themes that feature robots or other artificial human-like constructs or thinking machines with what has already been realized or is in the making (Fleck, 1984). One could re-analyze predictions that have been made in the past, sinister warnings as well as blissful prophecies regarding a cornucopeian future, and point to their inaccuracies, their faults in reasoning and their failure to grasp essential constants - but such an exercise, unless much more fully developed, would not necessarily guard us against committing similar errors today.

In fact, it was one of the premature and yet audacious ideas put forward by Otto Neurath, here in Vienna some 65 years ago, to work towards the science of utopistics (Neurath, 1979). What he meant was a kind of early technology assessment, with the crucial difference that it was not an isolated technological system or a singular technological development that was to be assessed with regard to its likely future consequences for society. Rather, utopian systems were to be systematically compared with each other, in order to detect the flaws in their reasoning or in their methods of extrapolation. They were to be tested in the usual scientific way, but in a kind of rigorously controlled thought experiment. For Otto Neurath, the great visionary of a new social and scientific order, who conceived of himself as a social engineer in the most noble connotation of this word, any utopistic scheme meant planning for the rational basis of societal life, according equal importance to our knowledge of its social, and of its scientific-technological foundations a program unachieved until this very day.

2. UTOPIAN AND DYSTOPIAN IMAGERIES OF THE PAST

In the absence of the development of such a science of utopistics, the imaginary constructions of ideal societies, including the place accorded in them to science and/or technology, or of parts thereof, can be analyzed in a historical mode (Elzinga and Jamison, 1984). The beginnings of the utopian imagination in Western Europe, were still modelled after the religious world-view of the times; it were spiritual and religious ideas that served as guide-posts for the more mundane programs of how to construct ideal societies on earth, while it was only with the advent of modern science and technology that secularization set in here as well (Manuel and Manuel, 1979). Technology, then as now, seemed to offer an easy way out of otherwise intricate social problems: it promised the

fulfillment of material wants for all, beginning with Francis Bacon, and
an end to human misery; aggression and wars would become superfluous and
even the daily disruptions and irritations would be eased away by technological efficiency (Bacon, 1627).

It was left to the ideologically responsive function of science to
take up the promise of consensus and harmony for a society full of internal strife and disorder, as was the case with England in the 17th century.
Utopian thought captured the imagination everywhere as an endeavour to
keep disorder at bay. In this function, it inevitably became loaded with
a surplus of order, both in a positive and negative sense, that it was not
able to shed ever since. In an age, in which turmoil and incoherence of
actual social life were palpably felt, utopian writings were idealizations
directed towards organizational and bureaucratic order. They prefigured
either liberal or authoritarian tendencies that led eventually to the rise
of the absolutist and the modern democratic nation-state. But once science had made accessible the "marvellous symmetry of the universe", its o-
rienting function was to transform rationality and celestial harmony into
the guiding vision for the architectural social structure to be implemented in this world: the cosmic perpetuum mobile became the model for a
social utopia which, once set into motion, was thought to function
perfectly forever, if and when similar universal laws, applicable to human
behavior were found (Winter, 1984). Utopia definitely ceased to be a
Christian-inspired heavenly Jerusalem and became a state which could be
brought about through action, guided by science, while connecting the idea
of scientific feasibility with universal happiness. The modern emerging
scientific enterprise was quickly turned into a rational as well as a
utopian vehicle, charged to bring about a social world constructed in its
mirror image. What else could pose as the unsurpassed master copy for a
social order to be built than the natural order with its display of
invariance, harmony and eternal laws?

But neither the permanent technological fix, nor the celestial
harmony inspired by scientific discoveries could in the end bring about
the realization of utopia NOW. The utopian horizon kept moving onwards,
not least because of the progress achieved by science and technology. In
the middle of the 18th century, the classical space utopia, governed by
rational, geometrical constructions in which the interests of the subjects
were held to be congruent with those of the social commonwealth, gave way
to the dynamic time utopia, in which a more open construction prevailed,
reflecting also a change in the conceptualization of time (Luhmann, 1980;
Koselleck, 1979; Nowotny, 1975). The role that science and technology
played in these transformations is not simply one of empirical inductions.
Rather, science and technology created a horizon for the myth of the
history of reason to unfold. The actual progress achieved provided the
empirical substance of verifiable experience, on the basis of which the
projection of the hypothetically possible occurred. Science and technology provided the methods, content and ideology to make a certain kind of
future thinkable.

It was a future deeply molded by the belief in progress. It became a
general rule for scientific and technological inventions to lead onto new
inventions which to predict precisely in advance was not possible, but
which gave new space for the utopian imagination as well as providing the
verifiable background for the belief in progress. Progress became the
dynamically stabilized difference between experience and expectation - not
yet tarnished by the shadows of its more negative side-effects (Koselleck,
1979).

One of the most salient characteristics of the scientific and technological optimism which radiates throughout the 19th century is its seem-

ing smoothness. It expressed itself equally forceful in utopian writings of the time. It was achieved, not through the miraculous workings of celestial harmony applied to fragile social constructions, nor through the ingenious congruence of individual and collective desires and actions, so characteristic of earlier utopias, but exclusively through the command that science and technology offered in reshaping social relations. Smooth functioning precludes, among other things, ordinary discontinuities as well as major catastrophes - an interesting gap already contained in Bacon's New Atlantis. In it, catastrophes are prevented from occurring in nature, because they have been subject to human control. Bacon, the fallen statesman, who knew well from his own life what catastrophes meant, eliminated them altogether from his vision of the future. Other utopias followed in the same vein. For the present generation, it is hard to imagine how arduous the belief in social happiness was that became the hallmark of the social utopias of the 19th century and the role played by science and technology in this scheme. Smoothness in operation as the guarantee for happiness, followed by a future built upon industrial work and, not surprisingly, social order was projected as functioning as easily as a well-run production plant. The direct line of descent of this notion can be traced right through to the enthusiasm with which the ideal of social planning was to be received in the early part of this century.

It was left to the rising dystopian vision, the correcting device for the excessive zeal of the utopian imagination, which would henceforth and irreversibly split the social order into those who controlled and those who were controlled, to bring into the open the underlying tension between utopia, conceived as an ideal societal construction, and science. As J.C.Davis has argued at great length, an inherent dilemma remains between utopian thought and scientific development, as long as science has the endless capacity for innovation and hence, for altering the conditions of social life as well. Utopia, in its desire to control and impose an ideal order, cannot tolerate in the end that which is fortuitous, spontaneous and which threatens to undermine its carefully constructed laws and ideals (Davis, 1984). Brought out in a bitterly satirizing or grossly exaggerating way in many dystopian examples, the social order is depicted as controlling every innovative, original, spontaneous act or thought since it threatens to undermine the order already established. Science, like falling in love, is accorded in this construct the subversive and dangerous potential for evading or circumventing established laws of social thought and conduct, by allowing the unexpected to happen. For this is the other side of the coin: while utopia and science share strong tendencies to reduce contingencies to laws and to build upon invariants, science alone, according to Davis, is an open-ended dynamic process, in which the unexpected, serendipidous and the accidental can still occur and are highly valued. The utopian imagination, Davis maintains, cannot possibly match the multitude of possibilities offered by science. The kinetic - moving - utopia is therefore a myth. It expects utopia to predict the course of future scientific innovation which, however, remains unpredictable in its core. But before proclaiming that utopia will either stop science or be overthrown by it, I suggest to examine what the rapid growth of AI and its applications have meant so far for utopian and dystopian thought and what new and unexpected twists the present argument might take.

3. AI: AN 'AUTOMATIC' END TO UTOPIAN THOUGHT?

(Note: this is the title with which James Fleck (1984) nicely captures the twist of the argument).

Ideas of artificial human beings or thinking machines have pervaded legend and literature from the earliest times. But it is only in the last 20 years or so, that technologies such as AI or industrial robots have appeared which seem to have the potential to realize these ideas. When formerly magical knowledge was seen as capable to produce artificial, human-like constructs, they stand today as symbols for scientific and technological advances in general. Yet, as Fleck shows, some of the underlying themes that kindle the utopian-dystopian imagination have remained surprisingly constant: the themes of robots as dangerous knowledge, and robots as projections of Man, of men and women, and of what being human is essentially all about.

It would lead too far to retrace here in detail the recurrent permutations of a few overriding themes that are to be found in the pertinent utopian-dystopian literature and above all in science fiction, the newly specialized branch of its more general literary predecessor. Fears and hopes seem to be triggered almost syncronically by the same developments, although dystopias are clearly gaining ground. When surveying the contradictory images that are thus created, one is struck by two observations: first, by the fact that the dominant themes and images, arguments and apparent refutations, are by no means confined to the literary domain alone, but are equally strong characteristics of the ongoing critical discussion on the social impact of the new technologies (Bjorn-Andersen et.al., 1982). Moreover, the utopian-dystopian line runs through the camp of practitioners as well as through the camp of their critics (one of the most prominent and early critics among the ranks of practitioners was Joseph Weizenbaum). While in the science fiction literature, for example, it is the survival of the human race, which is at stake, threatened to be overtaken by the artificial constructs that resemble them to perfection while topping them in efficiency and achievements, in the political discussion it is the extent to which machines will replace the human work force. In the literary genre, the difference which seperates true humans from their imitations has been treated in many permutations, emphasizing what is thought to be specifically human. It compares well with the ongoing debate on the possibly dehumanizing effects, once expert systems will be widely used, which again centers on what is thought to be the essence of human communications and interaction. Parallel warnings, for instance, are also raised recurrently with regard to the dangers of centralized control which technologically sophisticated systems facilitate, but also touch issues such as the preservation of cultural variety (Negrotti, 1984).

The other observation pertains to the utopian-dystopian dichotomy, so characteristic again of both, the literary-fictional and the actual, political discourse. It seems as though future developments function as an immense screen for the projection of present social interests and for the extrapolation of present hopes and fears. Exhortations and warnings oppose each other in rhetorics and argumentations, which can easily lead towards an eventually sterile debate. In summarizing attitudes towards thinking machines on the part of AI practitioners and outside critics, Fleck distinguishes between a simple utopian ideology of AI; the simple dystopian view of AI, which consists essentially in asserting its reductionist nature; and two more differentiated positions: one asserting that AI may be dehumanizing because it embodies an alien technological rationality, while the other view proposes that AI offers a way out, i.e. of humanizing technology, because it takes explicit account of human cognition. While this is a familiar controversy by now, its argumentative structure is by no means limited to AI alone.

Underlying this dichotomous mode of reasoning is a deeply embedded tension, which is inherent in the nature of the technology developped in

Western societies over the past 200 years, in consort with the social values that support it. As Langdon Winner has pointed out, this technology has always been most productive, when its ultimate range of results was neither foreseen, nor controlled (Winner, 1977). It always does more than intended. While this has been regarded in general as a welcome feature of technology up to now, since it serves as basis for the next round of ongoing developments, the unintended consequences are increasingly becoming more visible and questioned. The unstated common assumption until now was, that the positive consequences would - automatically ?- outweigh the negative unintended consequences. It is this unstated assumption, which seems no longer valid. We have to face the fact, finally, that in the forces that gave rise to the development of science and technology, unintended consequences were not _not_ intended (Winner, 1977).

Looking back to the early enthusiasm that AI and similar developments engendered, to "the days when everything seemed possible", and when one of the brightest dreams was the creation of a program that would mimic all human problem-solving abilities, we can clearly see that little thought was accorded to exploring the full range of second-order consequences. Even if today's assessments of the actual achievements, as Mitchell Waldrop puts it, both in the engineering camp, who are trying to get their programs to do smart things, and in the scientists' camp, who are after a general theory of intelligence, are much more modest. Waldrop maintains that the main thing that AI researchers have gained on the theoretical front is a certain humility, and of how much a computer has to know before it can do much of anything (Waldrop, 1984) but the vision of the revolutionary potential lingers on, both among the practitioners and among the general public, irrevocably interwoven with dystopian elements.

4. A NEW UTOPIA IN THE MAKING?

In the film I mentioned in the beginning, one of the dimensions lending credibility to the potential for survival of the Japanese society in an age dominated by microelectronics, computer technology and computational thinking, was the persistence of social rituals. Like all rituals, they serve to symbolize relationships, including those that connect human beings with "the spirit of things". It is one of the paradoxic and unanticipated consequences of the development that science and technology have taken, one of the evolutionary turns in the conceptual apparatus of societies, that this archaic notion of communicating with things, which modern science from the 17th century onwards has declared to be devoid of spirit and to be nothing but dead matter that can be controlled by the human mind, takes on new meaning and relevance in the complicated relationship between human beings and intelligent machines. By fusing "mere matter" with intelligence, by simulating, imitating, and by partly perfectioning the functioning of human reason, of thought and language operations, through such mediums as dedicated (!), massively parallel machines or of intelligent knowledge-based system architecture, an important shift is taking place. While previous scientific discoveries touched upon and re-defined the place of humans in the natural order, the present cultural environment has since long replaced it through its own artifacts. The crucial relation now becomes that of humans to their own creations - to their material products.

One way of re-defining this relationship consists in the exploration of the basis of consciousness, including consciousness that resides in and can be discovered as well as imputed into "mere matter" and artificial things. AI, in the eyes of many of its practitioners, contains a revolutionary potential which is based on the vision of a new epistemological approach, as yet only dimly understood.

From a relatively simple set of ideas and concepts, reflecting the prevailing dichotomous epistemology, work in and around this area is beginning to generate an enriched vocabulary and an artificial interpretative structure. This extended cognitive space, as it percolates into the wider ideological structure presently dominated by the dichotomous epistemology of man vs. machine, mind vs. matter, intuition vs. calculation, and subjective vs. objective, will ensure that the advent of robots and AI, no matter how limited or spectacular their capabilities, will be absorbed without either of the major simple utopian or dystopian outcomes being realized (Fleck, 1984). This new approach, which can be interpreted as the basis for nothing else but a new utopia, leads to the reintroduction of mind, albeit on a material basis. "It is believed that the new approach can accomodate contingency, chance and individual variability, without any intent to eliminate them. By challenging the man/machine and subjective/objective dichotomies, what is sought is not the extension of natural law to cover man, but rather, the elimination of a purely instrumental conception of science and the reintroduction of mind, albeit on a material basis, into the operation of the material world" (Fleck, 1984).

In one of her perceptive essays on the impacts of AI, Margret Boden examines expected progress in AI, both in core research areas that are likely to make rapid progress within the next decade and in what she calls the program of long-range AI research (Boden, 1984). In observing impacts of AI developments she notices foremost that AI will influence other sciences in their general philosophical approach as well as in their specific theoretical content. In her opinion, psychology and to a lesser degree biology have already been affected by computational ideas. While the behaviorists in particular had outlawed reference to mind and mental processes as unscientific and mystfying, AI, based as it is on the concept of representation, has rendered these concepts theoretically respectable again. Moreover, one should add, it has opened up a new and rapidly expanding field, called the cognitive sciences, and has led to the first, tentative formulation of what is called the Cognitive Paradigm (de Mey, 1982).

But the new relation to "the spirit of things" or to the mind embedded in matter is not only a visionary program for what is perhaps a new (utopian?) epistemology in the making. It is also to be found in social practice, here and now. Sherry Turkle has given a fascinating account of what she calls the "subjective computer" - the use of personal computers and the highly emotionally charged atmosphere in which users are working out their feelings of power and control, of being safe in a protected environment (Turkle, 1982). She suggests that the computer serves largely as a projective screen for other personal concerns. By many people it is experienced as an object betwixt and between, hard to classify and hard to pin down. She describes in the words of users how the elusiveness of computational processes, the tension between local simplicity and global complexity is experienced and contributes to making the computer an object of projective processes. In view of the computer's internal processes, individuals project their models of mind and in the descriptions given of the computer's powers, people express feelings about their own intellectual, social and political power - or their lack of it.

Thus, it is not surprising to encounter again some of the oldest anthropomorphic imageries, but also a yearning for security, for the possession of a safe corner of reality, amid another outside reality which offers it only to a small degree. The users described by Turkle are far from having an instrumental relationship with their computer, nor are they playful in the narrow sense of the word. Rather, they are very serious in wanting their computers to have a transparency that other things in their life do not have. The social world and the world created by science and technology seem to complement each other once more: the utopian pendulum can be observed in motion. And while it is easy to relapse into the uto-pian-dystopian mode of thought by interpreting the potential of the new relationships with personal computers as humanistic and hence beneficial, or by condemning them, by insinuating that, once they are widespread, they may become the new opiate of the masses, we should instead return to the embodiment of the new utopian pendulum, observing its swing between science and utopia a bit closer.

5. THE UTOPIAN PENDULUM IN MOTION

The mutual attraction and threat which science and utopia pose to each other in their common desire to subdue the contingent, has been described by J.C.Davis as the two horns of a dilemma (Davis, 1984): utopia can cope with science only - since science will inevitably change existing social arrangements and therefore threatens to destabilize them - when it conceives of a society that allows its members to control the moral and social consequences of scientific and technological discovery. This is a familiar dimension in utopian writings from the 18th century onwards until the present debates. Within such a utopian construct, the temptation is great to attribute fixity to science. In its extreme, the accidental aspect in scientific discovery would have to be removed, the spontaneous discovery harnessed in advance. Only then would it no longer menace the stability of the preconceived perfect social order, only then, presumably, would it be possible to extract only the beneficial yields of science and technology, while suppressing the negative ones.

The other side of the dilemma is the following: if science is not to be completely controlled and thus being reduced ultimately to a static and closed system, the ideal society has to be conceived as changing in a dynamic, evolutionary way. But can the utopian imagination really conceive of a continuous and endless sequence of legal, institutional and administrative devices, Davis asks, not only capable of adapting to successive changes, but also capable of guaranteeing their own transformation? Davis' answer is a clear no. Utopia will either stop science or be overthrown by it.

If it is impossible to foresee and to control all future consequences, intended and not intended ones, positive and negative ones, that will result from ongoing scientific and technological work, does it mean that a rampant technology has to be accepted? Put differently, are we stuck in the endless and sterile debates in the utopian-dystopian mode, until actual developments overtake the limits of the imagination by producing a much more differentiated pattern? For utopias and dystopias are always mirror-images of the societies that produce them; they are collective representations of the hopes and fears that these societies harbour with regard to a future that does not yet exist. Since utopian and dystopian thought are temporarily rooted in the present, they also tell us more about the present than about the actual future. In reading them as expressing the present oriented towards a future, and by observing and analyzing actual developments in their deviations from what has been hoped or feared, we are led eventually to a better understanding of how the future is actually made today.

For an observer of the contemporary scene, the future, once dreamt about in a paradisical or nightmarish way, has come to stay. While it is easy to be overly impressed by the scientific and technological forecasts that have been realized and have actually provided the islands with plenty and wishfulfilment, at least for that part of humanity that lives in the rich industrialized nations, its dark side has also come to stay with us. We have seen how the utopian-dystopian tension has moved along with the continuing debate about the social impact of science and technology's latest achievements, but we have not sufficiently appreciated the interaction between the social side of this development and the technological one. The apparent inability to synchronize rates of change, to adapt them to each other, to humanize technology and to invent new social rituals that will allow social beings to come to better terms with their own artificially created products, is the hidden message of the utopian-dystopian accounts. Whenever social problems are pressing, redemption is sought on the side of science and technology. Whenever their impact is perceived as potentially de-stabilizing, de-humanizing and threatening the social fabric, visions of a new society are created and their dystopian mirror-image signals an impending catastrophe.

Looking backwards, it is rather obvious that neither have science and technology been stopped by utopia, nor has science victoriously swept aside all utopian thinking. Quite on the contrary, utopia and dystopia have entered science and are here to stay. While it is impossible for the utopian imagination to anticipate or even keep pace with the actual developments of research and innovations from the outside, it has come to orient these developments from the inside. In doing so, - and discussions on AI and its impact are a good illustration - the utopian-dystopian tension is partly continued, but has partly been superseded by a new utopia: how to reconcile matter and mind, how to find the key to a new understanding of the universe in exploring the secrets of consciousness. The incorporation of utopia means also that the present becomes more and more loaded with choices. While science is seemingly producing a multitude of possible futures for our disposal, there can still be only one present. The hot fields in which present scientific utopias are taking material shape, show how a possible future is reduced to an instant present. The radical consequence to be drawn from this merger of science and utopia today is perhaps to realize that we are contributing ourselves to utopia and dystopia in the making and are confronted with having to live with them at the same time.

Ernst Bloch, one of the great writers on utopia and a utopian himself, wrote of the final stage: "es soll zu guter letzt, wenn keine Utopie mehr noetig ist, Sein wie Utopie sein" (in the end, when utopia is no longer necessary, to Be shall be like Utopia). Perhaps science has brought us closer than we ever imagined we would come, to the obligation of reconciling actual Being - the social side - with Utopia - the scientific and technological side.

REFERENCES

Bacon, F., 1627, New Atlantis.

Bjorn-Andersen, N., Earl, M., Holst, O., and Mumford, E., 1982, "Information Society: for richer, for poorer", North-Holland, Amsterdam.

Boden, M., 1984, Impacts of Artificial Intelligence, Futures, 2:60-70.

Davis, J.C., 1984, Science and Utopia: The History of a Dilemma, in: E. Mendelsohn and H. Nowotny, eds.

Elzinga, A., and Jamison, A., 1984, Making Dreams Come True - An Essay on the Role of Practical Utopias in Science, in: E. Mendelsohn and H. Nowotny, eds.

Fleck J., 1984, Artificial Intelligence and Industrial Robots: An Automatic End for Utopian Thought?, in: E.Mendelsohn and H.Nowotny, eds.

Koselleck, R., 1979, "Vergangene Zukunft", Suhrkamp, Frankfurt.

Luhmann, N., 1980, "Gesellschaftsstruktur und Semantik", Suhrkamp, Frankfurt.

Manuel, F.E., and Manuel, F.P., 1979, "Utopian Thought in the Western World", Harvard University Press, Cambridge, Mass.

Mendelsohn, E., and Nowotny, H., eds., 1984, "Nineteen Eighty-Four: Science between Utopia and Dystopia (Yearbook in the Sociology of the Sciences, Vol.8)", D.Reidel, Dordrecht.

de Mey, M., 1982, "The Cognitive Paradigm, Sociology of the Sciences Monographs", D.Reidel, Dordrecht.

Negrotti, M., 1984, Cultural Dynamics in the Diffusion of Informatics, Futures, 2:34-46.

Neurath, O., 1979, Die Utopie als gesellschaftstechnische Konstruktion, in: "Otto Neurath, Wissenschaftliche Weltauffassung und logischer Empirismus", R.Hegselmann, ed., Suhrkamp, Frankfurt.

Trappl, R., ed., 1984, "Cybernetics and Systems Research 2, Proceedings of the Seventh European Meeting on Cybernetics and Systems Research", North-Holland, Amsterdam.

Turkle, S., 1982, The Subjective Computer: A Study in the Psychology of Personal Computation, Social Studies of Science, 12:173-206.

Waldrop, M.M., 1984, The Necessity of Knowledge, Science, 223:1279-1282.

Winner, L., 1977, "Autonomous Technology", MIT-Press, Cambridge, Mass.

Winter, M., 1984, The Explosion of the Circle: Science and Negative Utopia, in: E.Mendelsohn and H.Nowotny, eds.

DISCUSSION

Nowotny: I think we have foreseen a discussion following this lecture, if
    I read the program correctly, and I would be very glad to answer your
    questions.

Francois: I am struck more and more with the growing difference between
    what indviduals really do and what they think they do. It is espe-
    cially visible with politicians in the whole world, but I think also
    with scientists. I wonder where society as a whole is going; and I
    wonder, if we are able to understand it.

Nowotny: Partly you have received an answer yesterday about the discrep-
    ancy between thought and action. However, I think this is part of
    social life in general but I am not so sure whether there is actually
    a discrepancy between thought and action. I think that our struc-
    tures are becoming more differentiated and we have a formal structure
    which is becoming more and more complex, because it has to deal with
    many issues. And then, we have informal local structures which are
    springing up and fill the vacuum that has been left and created by
    the growth of formal structure. And I think, if you want to close
    this discrepancy a little bit by wanting to become more honest and by
    facing reality, we have to look at this growing discrepancy between
    two kinds of structures, the formal and the informal or the hier-
    archical and the more decentralized structures. I think this is what
    you observe especially in political life and also in science. There
    has been an ongoing discussion that science presents one image of it-
    self as being a very formal rational system outside of public
    consumption, while internally there is knowledge about the informal
    ways in which scientists work; and this is also a discrepancy which
    is becoming more and more difficult to reconcile, because scientists
    are afraid that, if they were honest to the public about how actually
    scientific work proceedes, this would lead to a dramatic decline in
    credibility that science has in terms of its public standing.

From the floor: You have sketched the influence of artificial intelli-
    gence, at least in the future. On the other hand, we also see the
    opposite tendency. I think that you told us the more you know, the
    less you see how little you know. To some extent AI shows us how
    complex the human mind is and I see also tendencies of discrepancies.
    I know that there are people in AI who think that we get more and
    more knowledge. On the one hand this is true; on the other hand we
    see also the dissimilarity, the difference between machine and man.
    There is also a tension building up. It might be that we create an
    image of man as in La Mettrie's "L'homme machine."

Nowotny: I think what you describe is very true for the growth of scien-
    tific knowledge in general; I mean, the more we get to know, the
    more we have thinned out cognitive space and we discover what is ei-
    ther transcending it or what we have not covered. But I was nearly
    as much concerned with the social impact of this growth of knowledge
    and the growth of information. This is also one of the themes that
    is very dominant in actual discussions: How can - not scientists who
    do nothing but work on the growth of knowledge - but how can ordinary
    people cope with this apparent wealth of information and knowledge
    which is put at their disposal. The underlying problem is probably
    one of how to synchronize in a better way the growth of scientific
    knowledge and the growth of social structures that can absorb it.
    Now, we have not paid very much attention to this in the past,

because we were so fascinated by this tremendous progress that science and technology brought, and I have touched upon some of the fascination this has exerted. We are beginning to realize that we have neglected the social side, including the social capacities of absorbing this knowledge. I think this is really one of the most pressing problems that we are facing. So, I am not so much worried about the continuation of growth of scientific knowledge but how we absorb it.

From the floor: Your talk is about the Western European, the Japanese, the US perspective. But if you read about the main ingredients of the original American utopia - the utopia which is the United States - and that's the core, that's the key ingredient of their utopia, of freedom and free lives and a sort of primitive justice. You repeatedly noticed that most of the world, the larger part of the world is still without cars. Only one third of the world population can get a car sixty years after its inventionß;; and it is an everyday good to most of us who are here now. And if you think of the future of the technological progress, it might well be that progress is confined to the Western world, that is Western Europe, Japan, and the USA. The developing countries and the rest of the world, and also probably parts of Eastern Europe will be without those utopian ingredients in the future. So your use of 'we' is somehow dangerous because of these problems. And also in Western Europe it is not the case that we have only one utopia, the technological utopia, we have also had a wide variety of social utopias. We have social utopias and technologial utopias, and especially the social utopias have an infinite number of ends. We have the widest possible variety of different conceptions of justice and we have no idea how to put together all those ideas of justice to form one global consensus notion of a just world. And that is my second point. Let me repeat this: my first point is that it seems to be the case that we have to suppose that the technological progress will be limited to one part of the world, that's 'we'. And then we have two types of utopia, two different types of utopia. Once we have the technological utopia of being able to fulfill our needs by using some technical methods. And then, we have social utopia concerning our ideas of justice, and you should probably take into account the kind of double nature-utopia more carefully.

Nowotny: Well, I plead guilty of using "we" in a rather ethnocentric fashion; although in the very last part of my talk I stressed that what we have achieved is really limited to the rich, industrialized countries. With regard to your second point, there is, of course, a tremendous wealth of utopian writings, and the reason why I was asked to present this talk today was that I have edited a book, which will come out in the course of this year, on science in utopia. In it you find a number of contributions which take up some of the points that you have mentioned and which trace especially, how technology has been used in order to bring about social ideas of order, but where you can also see the faults in reasoning and how this technology becomes a kind of imperative that in the end subjugates again the social - so there is this tremendous richness in the relationship between social order and technological utopias and I have not been able to treat it in detail that probably is warranted. But I would like to stress again one point. I think we have to avoid reading utopias and dystopias too naively. They can be read as historical documents in order to tell us what people who lived earlier thought, how they imagined their future to be. But, I think we also have to see their function; and their function is really to elucidate something that is deeply problematic in the present. Utopias and dystopias are really a collective projection of this feeling about what is problematic. And if you look at the present utopian writings, there are almost no grandiose schemes in them, they have been abandoned. You have novels like Ecotopia or the science fiction literature which is a category in itself. What you find are fragmented utopias only, you have ideas on how certain social groups would live and create a future that takes care of their own needs. But the sort of grandiose design that was characteristic still in the 19th century has been shattered. Maybe it will return again when we have more courage to face these problems; but at the moment, I think, we do not have the imagination to deal with a global future.

From the floor: We cannot have any monolithic utopia in those grand schemes, because people are different and you cannot predict all of their decisions. We need a framework for utopia, where we can put in all our utopian ideas - everyone can have his own utopia, and the framework is where we put in all our utopian ideas; and that can accumulate a wide variety of different things. So we need a framework for utopia, not one single monolithic utopia.

Nowotny: In a way you can say that we have returned to a very individualistic utopia, a sort of micro-utopia. This happens, for instance, with regard to ideas we have about health; the whole sporting, jogging, mind-and-body market is one very individualized expression of such a micro-utopia.

Ghosal: What is the utopia today? Maybe a reality of tomorrow? Now my question is, how to depict social changes, and secondly, whether artificial intelligence can really help in predicting social changes. I am a little skeptic about that.

Nowotny: I share your skepticism. You know that about 10 years ago futurology in its different variants was a great theme and various institutes were founded and quantitative methods have been celebrated in advance that now finally would contain scientific rigor. "We are going to find out what the future will bring!" - until the oil-shock. This was a rather drastic break in this euphemistic view of predicting. And now, when one looks back, one sees that it is always easy to make predictions when you know that you are on an extrapolation line, and then you know what the future will look like. But with regard to discontinuities we are in a much more difficult position; however, as you know, the catastrophe-theory in mathematics has recently taken up this theme and we may eventually make some progress. But one lesson that the reading of many utopias and dystopias offers, is that there is this desire to subject everything that is accidental to control. In dystopias this takes the form of the repression of the accidental. It is a crime, if any accidental human action is committed - and falling in love is the literal device used very often to symbolize this accidental element. It is a crime that has to be repressed, because only then you can make successful predictions. I think this is an underlying dilemma that we face. If we do not only have the tools to foresee, but also the means of power to make happen what we predict, then I think we are living in a state that few of us would care to live in.

From the floor: What I miss in this discussion on utopia is the lecture of the minds here. We have a good argument to justify what is a good development and what is a bad development. If we ask: What is a better way to manage 'entropy vs. energy'? If I can make the same thing with a smaller amount of energy and a smaller amount of using entropy, I am on the right way. And in the biosphere we also see that there is a great necessity always to have together rules and chance. Another question about what you have mentioned, that is really a problem for our education: our children now learn in school scientific facts we heard not even at the university when we started, for the number of scientific facts has increased tremendously. But nearly everything has been eliminated to train our young people, how scientific researchers think and act. It would be much better to give our children the chance to develop an old mathematic thought by themselves than to feed them with a lot of mathematical techniques and to distress their ability to find out, how to think in the mathematic field.

Nowotny: Well, with regard to your first point of how to distinguish the good and the bad, or the positive and the negative features, I am somewhat skeptical whether there is the right way, especially when you want to apply it across societies or through different times. Although there are some "constants" of what is regarded as good and bad in every society, there is also a great variation. It is also difficult to discuss what is progress - in any sense that brings us further ahead, because the category of what is regarded as progress changes with what we have already achieved. So we have to take into account how the categories themselves - language and thought that express them - change with time and place and cultures. With regard to your second point I agree, I mean there is the tendency towards black-box-thinking, you are "presented" with something that is already designed and has the aura of being complete, and you do not know what goes on inside. So I can only hope that your children and my children will eventually become curious enough that they will also want to know what is inside the black box or what makes the black box tick.

From the floor: What do you think about the change in the relation between work-time and leisure time during the technological development?

Nowotny: Again, if I go back in the discussion: 15 years ago there was a lot of literature in the social sciences with special regard to leisure time. There was great concern, what people would do with their spare time, because technology would give us more spare time. This discussion has almost completely disappeared and the main concern today is really the issue of work and the place of work in the future of societies. Now, there is of course a regulatory mechanism underlying this problem that we associate with work. For work is not just a self-satisfying activity, but we use work for a distribution of income, we use it for a distribution of power and of other things society has got to distribute. It is this central conception and association - work that will become less in the future, because technology is taking over some of the things that now people are doing - which is the crucial discussion for me. Now, as you know, at least in Western Europe there are many discussions going on with regard to jobsharing and various schemes of flexible working time. You get the feeling that people know what they will do with their leisure time, that this is not really the problem. The problem is how to change the central distribution mechanism that hands out the good things, and especially income that people want, at the same time.

From the floor: From the very beginning of utopian thinking there has been the idea of harmony. I remind you of Johannes Kepler and Johann Wolfgang von Goethe. And this idea of harmony was mostly connected with "equilibrium"-thinking. In our present world we very rapidly need transition processes. Can we have harmony also included in rapid transition processes or should we wait until we reach another equilibrium?

Nowotny: I am not sure whether I can answer your question. You are very correct as to the importance of harmony. But this was also at a time where the harmony of the heavens was discovered, and this made a tremendous impression on people, to see that there were these laws that could be described - that there was harmony reigning in the universe. The shock of the state of disharmony and anarchy and political strife and aggression that existed in actual societies was a tremendous discrepancy. Now I think, we have perhaps reached a higher degree of ordering our conflicts, and I say this with great caution. After all, we all know the conflicts that we have not been able to order. But I think, we have made some progress in carrying out conflicts in a more orderly way. So this idea of harmony remains an idea. You suggested that there are transition periods in which conflicts are inevitable, and I tend to agree on that, because only then do old structures crumble and fall apart, and something new emerges; and this does not happen without conflicts. But I think there are also more rational ways of dealing with conflicts. So we should not accept conflict as something which is natural and inevitable, but that we also find ways of controlling conflicts - or at least, to go in this direction.

Hu: I would like to know what is your opinion of micro-utopia? Please give me more explanation on that.

Nowotny: What I called "micro-utopia" in this discussion here is an expression of the tendency towards a kind of hyper-individualism that you can find in Western societies. "Micro-utopia" means, in the extreme, that each individual wants to realize for himself or herself a utopia without concern to what others do. I think, therefore it is, in the end, an antisocial or potentially antisocial activity, while the earlier utopias always had the collective in mind and wanted to order the collective. Of course this was also more congruent to the type of thinking in the 16th and 17th century - where the individual was subject to the collective way.

Airaksinen: The terrrible mistake of all harmony is, that harmony is not very desirable in the long run, because it is so boring. We cannot tolerate harmony in the long run, because it drives us crazy.

Troncale: I am much in agreement with your description of the great distance between the technological advancement of society and its sociological responses. We seem to be very good at speeding up our technological advances, but there has been little matching activity on the sociological/values level to keep up with those advances. I submit that not only is the distance between the two large, but the rate of increase in the distance between the two is accelerating.

Nowotny: It is certainly true that this observation has been made before, but I am sorry that I also have to disappoint you - and I do not have the answer. And I don't think that anyone has the answer. We are only discovering now that it seems to be much more difficult to make progress in the social domain as compared to the technological domain, because we do not have the same extent of control in the social domain. We cannot experiment, as we are used to in the scientific, technological domain, with human beings, this is not possible. We cannot impose simply our ideas on others, we have to persuade them, we have to convince them, changes have to be brought about in indirect ways. One common underlying dimension which you also touched upon, is that of the rate of change. I think we would all agree that change is something that we want, and also that harmony is a state which is connected with the idea of fixity and statics, and this would give rise to a revolution very soon. But the rate of change and the management of the rate of change - temporal management - when to introduce changes, how to introduce them - can make us much more aware of the process nature of change rather than see technology as something which is "given", which comes as an external force; to start to perceive it as a process which is, of course, a socially constructed process. We still have a lot of work to do, but the direction in which I would go is to start thinking about the nature of the process including the temporal processes that lead to the introduction of technologies and their management. But it is not a satisfactory answer, I am sorry.

Troncale: I am surprised that the utopian thinkers do not incorporate "variability" in their systems. In natural systems, which have matured for about 13 to 20 billion years, everytime nature has come up with a new level of organization; she has always come up with a new mechanism of variation. This mechanism of variation is often inextricably linked to the mechanism of stability which existed on the previous level. One good example occurs in biosystems from the molecular level, to the cellular level, to the organ level, etc.. Every single level that emerged from the origins about 4.0 billion years ago - has always started off with a stabilizing mechanism for information, and then evolved a unique new mechanism for variation leading to the next level. Do utopian thinkers provide for this kind of change?

Nowotny: If I may add another commmment to that; it is one of the perennial and unsolved problems of writing in utopia to design a credible mechanism for change that keeps evolving. You have the expression of the "kinetic" utopia that "keeps moving on". But if you look at specific examples you can criticize them very easily. It is also the difficulty that we have in terms of our own social imagination - of imagining what these social mechanisms of change, of creating new mechanisms of variation would look like; retrospectively we can identify them, only the time scales are different for biological systems and social systems. For instance, the emergence of the nation state was a social invention , but it has taken centuries in order to get to this stage, and we are still working out some of the problems that come with the nation state. If you compare the biological evolution with that of social and political systems, you are, of course, in completely different time scales. Maybe 5000 years from now, if one looks backwards, you can say: "Well, in this time a new mechanism for handling change was evolving", and Stafford Beer gave a nice illustration yesterday of what may be evolving. But we may only know 5000 years from now.

If there are no other questions, I thank you for the very stimulating discussion.

THE PHYSICS OF COMPLEXITY

ASHBY MEMORIAL LECTURE

Robert Rosen

I was privileged to have known W.Ross Ashby personally, albeit briefly. We had the opportunity to interact rather intensively over a six-week
period in the mid-1960's, when we both participated in a Summer Colloquium
on Theoretical Biology, sponsored by NASA, which then had an interest in
such things. I have very vivid memories of those days, and of Ashby himself, and accordingly I am most honored to be invited to present this
Ashby Memorial Lecture.

What I propose to do is to critically review Ashby's ideas about the
brain, about biology, and about complexity in general, in the light of
some three decades of subsequent experience acquired since the first publication of Ashby's two great books, Design for a Brain (1952) and Introduction to Cybernetics (1956). It is relatively easy to do this, since
Ashby, like Waddington and many other English theorists, had an enormous
gift for writing lucidly and explicitly about even the most complicated
matters. Thanks to this crystalline style, often lacking in other writers
(including myself, I am afraid), one always knew where Ashby stood, and
exactly what was being assumed at each stage of any discussion.

Ashby's general approach to biological problems was not reductionistic, but it was Cartesian and Newtonian. That is to say, Ashby was a
mechanist. Indeed, in many ways he represented a kind of culmination of
the mechanistic approach to organisms and their behaviors; it is precisely for this reason that one can learn so much from him.

His general approach to problems of biological organization was set
down many times, but never more clearly than in his book, Design for a
Brain. The problem which Ashby set himself was set forth at the outset
with his customary clarity:

I hope to show that a system can be both mechanistic in nature and
yet produce behavior that is adaptive. I hope to show that the essential
difference between the brain and any machine yet made is that the brain
makes extensive use of a method hitherto little used in machines. I hope
to show that by the use of this method a machine's behavior may be made as
adaptive as we please, and that the method may be capable of explaining
even the adaptiveness of Man.

To pursue this problem, we must first characterize what we mean by
"mechanistic", and how a mechanistic object or machine is to be studied.
For Ashby, the concept of "machine" is co-extensive with that of "material
system", or with what I myself have called a "natural system". It could

be an atom or an organism or an ecosystem or an automobile; the only requirement is that it populates the external world of events, rather than
the internal, subjective world of ideas, impressions, and symbols.

Such a "machine" is to be studied objectively through real or idealized processes of measurement. Such measurement processes lead naturally
to the idea of what Ashby calls a "variable" (i.e. what are now generally
called "observables"). He defines this as follows:

A variable is a measurable quantity which at every instant has a def-
inite numerical value.

But, as Ashby recognizes, every real machine presents us at the outset with an infinity of such variables. We cannot directly study such an
infinity; thus to study any machine through its variables and forget a-
bout all the rest, i.e. forget about all of the machine. The result of
our choice, then, is to create an abstract object, consisting of a finite
number of variables associated with a machine; it is such an abstract object which Ashby calls a <u>system</u>.

A <u>state</u> of such an abstract system is a set of numbers; namely, the
values which all of the system's variables assume at a particular instant.
The behavior of the original machine (i.e. its temporal sequence of e-
vents) will thus reflect itself as a sequence of state transitions in any
abstract system we create by concentrating on any finite set of its variables. But, as Ashby points out, we can now identify a special subclass of
systems within the infinitude of ways of selecting a finite set of
variables out of the original infinity which a machine presents to us.
This is the subclass of what Ashby calls <u>state-determined systems</u>; finite
sets of variables for which, at any instant, the state transition is
completely determined by the present state. In fact, he makes the following explicit postulate:

Given a (finite) set of variables (i.e. a system), we can always
find a larger (finite) set that (1) includes the given variables, and (2)
is state-determined.

Ashby remarks that "the assumption that such a larger set exists is
implicit in almost all science, but, being fundamental, is seldom mentioned explicitly".

Thus, of all the infinity of abstract images of machines (i.e. systems), we are most interested in the state-determined ones, which Ashby
points out, share with the machine itself the property that "if its internal state is known, and its surrounding conditions, then its behavior follows necessarily". And of the state-determined systems, we are most interested in those which are <u>simplest</u> in some sense.

Thus Ashby posits quite a string of abstractions; from machine to
system, to state-determined system, to simplest state-determined system.
He asks, rhetorically, why the study of such abstract things should be of
value for biology, with its enormous complexity and variability. His
answer is central to his entire scientific enterprise, and has two interrelated facets: (a) such abstract systems can be studied <u>precisely and
exactly</u>, so that in principle they can be completely understood; (b) less
abstract systems, and ultimately the machine itself, while not corresponding exactly to the systems we have studied, are nevertheless <u>close</u> to
one or another of them. In Ashby's own words:

(We) must try to be exact in certain selected cases, these cases being selected because we can be exact. With these exact cases known, we

can then face the multitudinous cases that do not quite correspond, using the rule that if we are satisfied there is some <u>continuity</u> in the systems' properties, then insofar as each is <u>near</u> some exact case, so will its properties be <u>near</u> to those shown by the exact case. (Emphasis added.)

This idea is really the crux of Ashby's mechanistic epistemology, and we shall return to it a number of times as we proceed.

In concluding this brief review of Ashby's ideas, we must mention explicitly that Ashby viewed his mechanistic approach as the only valid scientific alternative to teleology. Indeed, he regarded teleology as fundamentally antithetical to true science. Thus, he says:

It will be assumed throughout that a machine or an animal behaved in a certain way at a certain moment because its physical and chemical nature at that moment allowed it no other action... our purpose is to explain the origin of behavior which <u>appears</u> to be teleologically directed.

In other words, by showing explicitly how a mechanism can manifest apparently telic behavior, Ashby wished to show that concepts like "goals" or "ends" were at best superfluous and at worst mystical and unscientific.

What I propose to do now is to briefly indicate how the mechanism assumed by Ashby represents a direct embodiment of 17th century Newtonian mechanics. These Newtonian ideas, and the epistemology underlying them, have permeated all of our ideas about systems and their behaviors ever since; in fact, they are tacitly assumed to be the <u>only</u> way that systems can be studied. However, by precisely isolating these epistemological presuppositions, it is possible to see explicitly that alternatives indeed exist; to make a case that the Newtonian picture is in fact unduly restrictive, and must be extensively modified if we are to progress.

It must be clearly recognized at the outset that the influence of Newtonian mechanics has radiated in two main directions; a reductionistic direction and a paradigmatic direction. The former argues that, insofar as every material system can be regarded as a system of mass points, the mechanics of Newton (or some extension, like quantum theory) in principle contain the solution of every scientific problem. All we need to do to understand <u>any</u> material system is to characterize its particles and the forces acting on them, formulate the necessary equations of motion, and integrate them. Ashby himself does not embrace its paradigmatic aspect; that the <u>language</u> in which Newton described his theory of systems of mass points is the universal language for talking about systems in general, even if they have not been, or cannot be, reduced to systems of mass points. Indeed, Ashby's state-determined systems are nothing but a paraphrase of the Newtonian <u>language</u>, adapted to inherently non-mechanical situations. The essence of this language, as we shall see, is that systems have states, and that their behaviors are represented by dynamical laws superimposed on these states. These states are the cognates of mechanical phases; (more precisely, of impressed forces). In one form or another, every mode of system description known to me is a technical adaptation or modification of this basic presupposition.

The Newtonian ideas were regarded in their time as the supreme embodiment of the concept of Natural Law. Thus, we must digress for a moment to discuss this concept.

The idea of Natural Law has two quite separate facets. On the one hand, there is implicit in it a belief that the sequence of events manifested in the external world is not utterly capricious or arbitrary or chaotic, but rather that there exists some <u>relation</u> between them. The relation between events in the external world can be summed up in a single word: <u>causality</u>. Thus, the first facet of a belief in natural law consists of a belief in a <u>causal order</u> relating events we perceive in the external world. We could not do science, and in fact we probably could not stay sane, without a belief in causal order.

But that is only one facet of our belief in Natural Law. The other one is that this causal order relating events can be (at least in part) grasped and articulated by the human mind. This means ultimately that the causal order relating events can be translated or mirrored by corresponding relations between <u>propositions</u> describing these events. But propositions are mental constructs of a linguistic, symbolic character; relations between propositions cannot be causal. Nevertheless, there does exist a relation between propositions, playing the same role as causality does in the external world; that relation is the logical one of <u>implication</u>. Thus, the other half of our belief in Natural Law is this: that the causal order relating events in the external world can be imaged by implications between propositions describing these events. Indeed, I would argue that the whole task of theoretical science is to bring causal order into congruence with implicative order within an appropriately constructed formal image.

The formal images of causal order belong, in the broadest sense, to mathematics. I would argue that mathematics is nothing but the study of implication in formal systems; it is the art of extracting conclusions (theorems) from premises (hypotheses). When we have properly brought such a mathematical system into congruence with some causal structure in the external world, the theorems of that system thereby become <u>predictions</u> about the causal order.

A great deal of theoretical science is concerned with characterizing the class of mathematical systems which can be images of causal structures in the external world. One of the achievements of Newtonian mechanics was to posit a class (or in mathematical language, a category) of such formal images; a category of dynamical systems. In Ashby's language, this is essentially the category of "state-determined systems", and as has already been noted, this has been the arena for all of systems theory ever since. Many of the deep problems of theoretical science deal precisely with this category of mathematical images, and the mathematical relations which exist between them; the problem of reductionism, for instance, involves nothing else.

Now let us return to the concept of the causal order between events, which as we have argued is one essential part of our belief in Natural Law. We may first note that, as a result of the pervasive belief in the universality of the Newtonian ideas, many perceptive scientists and philosophers (including Bertrand Russel) have argued that the very notion of causality is obsolete and pre-scientific, and should be expunged from science. These writers have noted that the word "cause" does not appear any more as a technical term in mechanics (or in physics in general), and that <u>therefore</u> it has no meaning. This position only reflects the complete tacit acceptance of the view that the order between events has already been completely imaged in a formal category of "state-determined systems", and henceforth we need not concern ourselves further with the imaging process itself, nor even, for that matter, with the events themselves. And since causality pertains to relations between events, and not between propositions, it does indeed disappear from explicit view when we forget about events and only consider their formal images. But this "forgetting the events" is itself a process of abstraction, and as we shall see, we do it at our peril.

The first, and still the most influential, treatment of the concept of cause as a relation between events was provided two millenia ago by Aristotle. Aristotle was the only one of the great philosophers who was primarily a biologist, and this fact colored his thinking in a unique way. Let us briefly review what he argued. In his view, the entire business of science was to grasp "the why of things"; since the answer to "why" is "because", he was thus naturally led to consider the notion of cause in terms of the ways of answering the question "why"? In a nutshell, he argued that there were four distinct and inequivalent ways of saying "because", and these led him to posit four corresponding categories of causation. In modern parlance, these are: (a) material causation, which roughly has to do with the physical basis of an event ; (b) formal cause, which concerns what we would now call program; (c) efficient cause, which we should now call a program-determined operator on material cause; (d) final cause, which concerns telos or end. For over a millenium, Aristotele's views dominated what there was of science; science was the study of causes. This situation persisted until Newton replaced it with a return to even older views of the pre-Socratic Greek atomists.

These categories of causation, or at least most of them, do have mathematical images in the Newtonian picture. For instance, if we regard "the state of a system at time t" as an effect, and its mathematical image as obtained from integrating the equations of motion, then material cause translates into initial conditions; formal cause translates into structural or constitutive parameters; efficient cause translates into the integral operator which generalizes what the engineers call transfer function. But there is no final cause in this picture. Indeed, as Ashby noted, the concept of final cause would have to involve a notion of the future acting on the present; of future state or input affecting present change of state; of anticipation. This is resolutely excluded, once we have decided that the category of dynamical (i.e. "state-determined") systems constitutes the only acceptable class of mathematical images of the external world. Since Newton's time, this assumption has been made automatically, and it is essentially for this reason that telos, and its associated notion of anticipation, have been routinely excluded from science. The Newtonian picture we have adopted simply cannot accommodate them and survive.

Let us look again at the relation between the Newtonian picture and the Aristotelian categories of causation. The essential point is that, in the Newtonian picture, these categories are isolated into independent mathematical elements of the total dynamics. For instance, if "initial state" is identified with "material cause", then the concept of state space segregates the category of material causation from the other categories, and enables us to manipulate material causation while leaving all the other categories of causation unaffected. Likewise the formal cause, which is segregated into a parameter space, and with efficient cause, segregated into either an input-dependent family of integral operators, or in differential form, into the dynamical laws themselves.

Using these ideas, it can be shown rigorously that the Newtonian language, which we have accepted tacitly and uncritically from the outset as the universal vehicle for system description, is equivalent to asserting that the categories of causation are entirely isolated from each other; that we can modify any one of them separately, leaving everything else fixed. When looked at in this light, perhaps that language does not look quite so universal, after all.

Even if this independence of causal categories is accepted, however, the categories of causation are still inequivalent. This means, in more precise terms, that e.g. the same effect cannot be produced by a variation in initial conditions alone (material cause), and by a variation in constitutive parameters alone (formal cause), and by a variation in environmental controls (material cause) alone. Or, what is the same thing, that a variation in one category of causation cannot be offset by corresponding variations in the others. The problem of determining under what circumstances the categories of causation are equivalent in this sense translates mathematically into a problem of stability of mappings in parameterized families, and as is well known, not all such families can be stable; there will in general occur bifurcations. Indeed, bifurcations must occur whenever we compare an equivalence relation (similarity of behavior) with a topology (nearness of parameters). But this is precisely the kind of situation which arises with Ashby's assertion about "continuity of system properties", based on the argument that any system will be "near" a simple state-determined one. Thus, even if we accept the Newtonian language, with its segregation of the categories of causation into independent mathematical structures, the inequivalence of these categories raises crucial theoretical questions which have never really been addressed. And indeed, insofar as to be "simple" is non-generic, we may expect as a general rule that bifurcations will occur precisely around these "simple" systems, which we are attempting to use as models for all systems. In the cases which Ashby cites to justify this whole approach (ideal gases, frictionless oscillators, etc.) this is exactly what happens; closed , isolated, conservative systems and the like are inherently so degenerate and nongeneric that literally anything can happen when we open them up.

The infatuation of contemporary physics for this kind of degeneracy and non-genericity goes quite a long way in explaining the scandalous absence on any important relation between even the most powerful theories of physics and the most marginal biological phenomena. Indeed, viewed in this light, it is most ironic that theoretical physics should fancy itself as concerned with universal laws, and in quest for these should have disdained biology as dealing merely with an insignificant class of inordinately specialized systems from which no universal principles could possibly be expected. And doubly ironic is the abject acquiescence of many molecular biologists in this view, seeking to bury themselves and their field in a specious reflected association with remote and inapplicable universal laws. In fact, the situation is quite the reverse; contemporary physics is not the general nor biology the particular. Indeed, if physics is ever to become in fact what it presently claims to be, namely the science of material nature in all of its manifestations, then it must come to terms with the realities of biology, and in doing so will be forced to transform itself out of all present recognition. Some slight inkling of what will be involved in this can already be seen through contemplation of what the concept of the "open system" has done to thermodynamics; where after nearly half a century there is still no physics capable of dealing with even the most rudimentary biological (or even physical) situations. But that is another story.

Now let us return to the main line of the argument, and look briefly at what happens when we abandon the requirement that the categories of causation must be represented in independent mathematical structures. This means, in Ashby's terminology, that we give up the idea that each "variable" of a system can be classified as belonging exclusively to the category of material cause, or exclusively to the category of formal cause, or exclusively to the category of efficient cause. In other words, we allow that some, and perhaps all, variables simultaneously participate in two or more of these causal categories. Then what happens?

What happens, of course, is that we must allow a wider class of mathematical images of physical reality than the dynamical systems, or "state-determined" systems, to which system theory has hitherto restricted itself. In this new mathematical world, it generally happens that the value of a system observable, the value of its rate of change, etc., are independently determined, instead of all being derivable from a single rate law as in the Newtonian picture. These mathematical images become more like webs of informational interactions, no level of which can be derived from any of the others. In particular, there is no "state space" which can be fixed once and for all. The class of all of these new mathematical images form a category, in which the category of "state-determined" systems sits as a very small subcategory, just as the rational numbers sit as a set of measure zero in the set of all real numbers. However, just as in that case, it turns out that the members of the big new category can be regarded as limits of sequences of dynamical systems. That is, there is a sense in which the behavior of one of these webs can be approximated, albeit only locally and temporarily, by an appropriate "state-determined" system. So there is still a notion of approximability, but it is very different from the one Ashby visualized so long ago. The fact that the new approximability is only local and temporary explains a great deal about why we have been able to go as far as we have with the non-generic Newtonian picture, and why we have never been able to go further with it. The situation is similar to that faced by the early cartographers, who were attempting to map a sphere with pieces of planes; here, the Newtonian language should be thought of as the planes, and the new images, of layers of independent informational structures, as the spheres. Locally, the difference between sphere and plane disappears, but as we attempt to map out larger and larger regions on the sphere, we have to keep changing our planes. The sphere is in some sense a limit of the local planar pieces, but these pieces are related by a global condition (i.e. the topology of the sphere) which cannot be found locally. And the requirement that we must continually pass to other planes as we attempt to map more and more remote regions can, depending on how we look at it, be regarded as error (the discrepancy between planar and spherical surface) or as emergence (of the curvature of the sphere).

I have elsewhere proposed that this new category of presumptive mathematical images of physical reality be called a category of complex systems, while the subcategory of "state-determined" systems be called the category of simple systems or mechanisms. There are many reasons for choosing this terminology; among them, it is a corollary of their structure that a complex system, in the above sense, possesses a multitude of simple system descriptions, which cannot be combined into a single "master description" of this type. I had earlier taken this to be the very definition of complexity.

Viewed in this light, then, physics and all of its manifold system-theoretic variants comprise a science of simple systems. And organisms are not simple systems. Thus, I can visualize a science of complex systems, from which both contemporary physics and biology, in two distinct ways, emanate.

In closing, let me indicate one corollary of passing to the more general framework of what I have called complex systems. Namely, by loosening the Newtonian shackles, we can introduce a category of final causation in a perfectly respectable, non-mystical way. In other words, the concept of anticipation is meaningful in the category of complex systems. This fact alone, perhaps, is sufficient justification for looking seriously at this world.

We have thus come a long way from the world of mechanisms which Ashby studied so long and so thoroughly. I am sure that he would be aghast at much of what I have said, but I am equally sure that he would take it seriously. And indeed, much of my motivation for probing beyond the limits of the Newtonian paradigm arises from my knowledge of Ashby and of his work; had the problems with which he dealt been solvable within the paradigm he was using, he would have surely solved them.

KNOWING NATURAL SYSTEMS ENABLES BETTER DESIGN OF MAN-MADE SYSTEMS:

THE LINKAGE PROPOSITION MODEL

Len R. Troncale

1.0 INTRODUCTION

1.1 The Principles of Systems Science as Guidelines for Human Systems Engineering

How does this article fit into the set of loosely related articles and the umbrella title of this book? The Linkage Proposition Template Model (hereafter LPTM) has some interesting relationships with the future of words such as Power, Autonomy, and Utopia, as well as to understanding the inner workings of the complex systems found in nature.

First, the concept of power. Bacon said "knowledge is power". The history of human civilization supports the statement. From such examples as the advent of manipulation of symbols, predictions of star movements, and the building of boats by early civilizations, to the influence of Aristotle on Alexander the Great and the design of war machines by Leonardo da Vinci, the world has witnessed "special knowledge" as a tool for the powerful. At no other time in history has it been so clear as today: scientific and technological "special knowledge" is used both in peace and war, both by nations and individuals to amass and sustain power. Unfortunately, knowledge is not wisdom, and so power from knowledge is u-tilized in questionable ways. In this and other papers (Troncale, 1978, 1982b, 1984a, Troncale and Voorhees, 1983), I present the case for a new "spec knowledge" called systems science, and for a specific, and highly particular case of systems science, the Linkage Proposition Template Model (LPTM). It is allied to the work of virtually all the authors of this volume who regularly report their findings at systems meetings. Here I will argue that the LPTM is potentially a practical tool and a pathway to power for good or ill. Elsewhere, I argue that systems science studied appropriately has the potential for joining wisdom with its inherent spe-cial knowledge, such that the new source of power may be utilized more wisely (Troncale, 1984b).

Next, the concept of autonomy. The mechanisms included in the LPTM are prototypical descriptions of the detailed interactions between proc-esses found in most "mature" systems. In systems science, the "processes" are called "isomorphies" and the interactions among the isomorphies are called "linkage propositions." Hypothetically, these specific mechanisms are the cause of the higher level functions of systems, such as systems origins, form, maintenance, flows, growth, development, transformations, couplings, death/decay, field characteristics, evolution, and emergence (Table 2; all tables can be found at the end of the paper). It is by the action of these higher level functions that long-term stable systems appear in nature: where stable here means that they can be observed by man. Systems can be observed because they do experience origins followed by a complex series of control functions that lead to their stabilization, which upon analysis is a description of their autonomy. The many iso-morphies and linkage propositions of the LPTM can be used, therefore, to explain the higher level function of systems autonomy. The LPTM is an ap-proach to understanding autonomy on a general systems level that is an al-ternative to the approaches explained in the second part of this book.

And finally, utopia. At the present time the explanations and under-standing derived from systems science are so primitive that at least two generations of hard work face us before we can begin to claim demonstrable "special knowledge" at a level acceptable to the most critical minds of other disciplines. In a recent paper (Troncale, 1985a), I cite no less than 33 obstacles to significant progress in the field. Still, several small, but dedicated professional societies, composed of workers trained in virtually every field known to man, are sufficiently convinced of the potential of this infant area of research that they are willing to expend valuable time and resources on its behalf. It will take a considerable leap of imagination, then, for anyone to suggest seriously that systems science, and the LPTM in particluar, may promise to bring humankind closer to the historical, perhaps apocryphal dream of utopia. In any case, this author declines to support any of the visions of utopia from More's to the present, not only because one man's utopia is another's hell, but because systems science studies indicate that the overly stable system often envi-sioned by utopian plans is inevitably doomed to change, or at least would be challenged to evolution and emergence so compellingly, and by forces and needs so unforeseeable, that only a very general design could be suf-ficiently flexible to survive. But wait. Isn't that the type of design that may eventually be expected to emerge from systems science? Consider the following.

This paper will suggest how the LPTM could be used to guide several generations of research into the mechanics and holistics of natural sys-tems functions. These natural systems have had more than 13 billion years to equilibrate and optimize in an environment which would allow the sur-vival of only the most optimal behaviors within the contexts of each other. No human system exists outside of the very same contexts and envi-ronment that natural systems have adapted to by necessity. So presumably the detailed study of these surviving, "best-case" interactions that ex-plain "systemness" will include many usable and valuable guidelines for humans in their heretofore rather blind engineering of their own systems.

We need such science-based guidelines desperately. While "repro-ducible" and "cumulative" progress on science and technological assistance to the human condition has improved markedly over the last 400 years, "re-producible" and "cumulative" progress on man's operating values and treat-ment of other human beings has not kept pace, even though attempts have been made for at least 4000 years. As far as we presently know, man can-not command laws outside of those currently acting upon nature as so elo-quently pointed out by Bronowski (Bronowski, 1978) in reaction to a recent resurgence in interest in forces outside the conventional. Science-based laws coupled with conventional wisdom are all that's reliably and reproducibly available. So the kind of "special knowledge" inherent in the LPTM, and systems science in general, may be viewed (tempered by some healthy scepticism) as not only a source of knowledge and its subsequent power, not only as a stunningly detailed glimpse of autonomy-inducing processes, but also, in the very long run, as a blueprint for a reasonably limited and modest utopia.

1.2 The Limits of General Systems Science: The Need for Linkage Proposi-
tions

The above statements mark the hope and the potential.   What is the
reality of systems science today?

It is tough betting your life on the emergence of a new field, especially one faced with as many paradoxical demands as general systems
science. The systems scientist attempts to be universal enough to capture
the holistic nature of systems in an environment totally dominated by reductionists. The isomorphs he seeks to discover and elaborate are universal by definition. Yet, simultaneously, he must seek particularity if he
is to achieve the fulcrum, the handle on reality, needed to move reality.
Kepler said, "to measure is to know." If the universal knowledge of the
systems scientist is too general to have demonstrable correspondence with
practical reality, and consequently some degree of measurability and re-
producibility, the world judges that he has no "special knowledge" at all.
He must achieve both universality and particularity, analysis and synthesis at the same time, while surrounded by intelligent beings who support
one extreme or the other, and who further insist that the two approaches
are inextricably opposed and that everyone in the "other" camp is a fool.

The historical answer is to create a new technique that transcends
the paradox and convinces the skeptics. Genes were not recognized to exist in pieces until the discovery of restriction enzymes and rapid nucleic
acid sequencing techniques. The continental drift theory and plate tec-
tonics had to wait for the invention of paleomagneticism before consensus
could be reached. The techniques of bifurcation and catastrophe theory
and cobordism surgery in topological math had to be elaborated to give
substance to models of discontinuous change. Usually the newly-invented
technique overcomes serious obstacles inherent in the field of study.
Systems science still awaits the discovery or invention of a tool or technique that overcomes the obstacle of paradoxical demands for universality
plus particularity. Yet natural systems accomplish simultaneous universality plus particularity effortlessly. Perhaps the LPTM in modeling natural systems may be a step in the direction of the needed transcendent
tool.

In fact, the Linkage Proposition Template Model is designed to overcome some of the obstacles facing general systems science (Troncale,
1985a). Systems science has been rightly criticized for:

  * Not Using the Full, Minimal Set of Isomorphies.

  * Not Adequately Studying Linkages Between Isomorphies.

  * Not Adequately Describing the Self-Generating Nature of the Full,
    Minimal Set of Isomorphies. (Self-induction or autonomy).

  * Not Achieving an Adequate Taxonomy of Isomorphies and Related
    Systems Types.

  * Not Adequately Relating Isomorphies and Their Interconnections to
    Fundamental Systems Functions.

These shortcomings result in a loss of the particularity necessary to
explain how systems come to exist in nature. All of these obstacles must
be overcome before systems science can begin to argue that it possesses
the "special knowledge" we have come to expect of a new field, and before
it could be used to guide the design of human systems. The LPTM is one of
several projects that attempts a direct answer to the challenges presented

by these obstacles. It does this by identifying many isomorphies and expressing their even greater number of interactions in detail. Before presenting these "linkages", it will be necessary to explain what is meant by the term "isomorphy", and to describe the sources of isomorphy.

2.0 ISOMORPHIES: FOCUSING ON THE SPECIAL KNOWLEDGE OF SYSTEMS SCIENCE

The field of systems science is polymorphic, containing the sub-fields of systems analysis, systems theory, and general systems theory. These subfields are quite dispersed along the selfsame spectrum of systems studies, ranging from applied to theoretical research extremes, respectively (Troncale, 1982a). Consequently, the identification of its "special knowledge" is debatable. Table 1 shows some distinctions among the sub-fields cited. In fact, most critics of the field (Berlinski, 1976, Boffey, 1967, Majone and Quade, 1984, Hoos, 1983) often ignore these distinctions and attack the completeness of the methods of systems analysis, or the decades-old starting assumptions of general systems theory, as Klir points out in a review of Berlinsky (1976, American Scientist). Table 1 indicates that isomorphies have different roles in each subfield and so the "special knowledge" of each differs. One cannot fault the critics, if the field itself ignores these distinctions; an omission which creates confusion, inappropriate expectations even within the field, and increases the difficulty of intergration and cumulative progress across the three sub-fields. So the LPTM attempts to be clear in its descriptions of what the most fundamental "special knowledge" of systems science is, what its limits are, and in presenting specific, addressable criteria on what should be excluded from and included in its corpus of knowledge. The LPTM papers (Troncale, 1978, 1982b, 1984a, Troncale and Voorhees, 1983) emphasize that most of the tools of systems analysis, and most of the explanatory power of systems theory derives from isomorphies and that these are, therefore, the proper focus of study of general systems.

Historically, isomorphies have been discovered or elucidated by individuals working at all levels of inquiry into nature, including many from the conventional disciplinary specialties:

* mathematicians (Mandelbrot, 1977, Thom, 1975, Wiener, 1948),

* biologists (von Bertalanffy, 1968, Maturana and Varela, 1980, Rosen, 1970, Waddington, 1977),

* chemists (Eigen and Schuster, 1979),

* astronomers (Whyte, A.Wilson, and D.Wilson, 1969),

* physicists (Enns et.al., 1980, Haken, 1983, Prigogine, 1980, Zabusky and Kruskal, 1965),

* man-made systems engineers (Iberall, 1972, Klir, 1969, von Foerster, 1974, Warfield, 1976a),

* sociologists (Miller, 1978, Parsons, 1971),

* management scientists (Ackoff, 1971, Ashby, 1963, Beer, 1972, Checkland, 1981),

* economists (Boulding, 1978, Simon, 1969),

* political scientists (Churchman, 1968, Deutsch, 1966, Easton, 1965,

Rapoport, 1968), sometimes even by

* philosophers and ideologues (Bunge, 1959, 1972, Jantsch, 1980, 1981, Koestler and Smythies, 1969).

But the future empirical refinement of these isomorphies and especially the systems consequences of their interconnection may best occur within the boundary of a new reconceptualization of general systems science.

2.1 Isomorphies Are Real: Part One. (Part Two in Section 5.1.1)

Isomorphy is a term that was used in the fields of mathematics and physics long before it was adopted in the fifties by the founders of systems science. In the former fields it drew attention to the existence of similarities in equations or parts of formulae used to model or describe different phenomena. Literally, the two disparate phenomena shared similar ("iso") form ("morph"). The use of the term in systems science is less rigorous and more encompassing. Similarities of process, structure, behavior, and effect, whether expressed mathematically or not, are described as isomorphic. The application domain is also extended. Isomorphs in general systems science are expected to be true across the full span of knowable, mature systems, although the limited state of current knowledge accepts provisional isomorphs that to date have been shown true for only a few systems.

Isomorphs are the fundamental level of information in general systems science. All else is built up from the level of isomorphy. All else is philosophy, design of practical tools, epistemology, application, or description of methodology. Isomorphs are the principle systems concepts, the "special knowledge" of the field, the theoretical basis.

Although these are declarative statements, the field itself is far from reaching consensus on the fundamental nature of isomorphies. In fact, the word is used differently by different practitioners. The most common usage relegates isomorphy to a simple comparison between two real systems. In this usage, which is similar to its usage in physics and mathematics, isomorphs have no reality of their own. They are the result of conscious mental comparisons carried on in the relatively artificial world of the human mind. Especially in the less rigorous usage of the systems field, they are relegated to the status of analogy and metaphor, hardly a respectable caste.

An emerging conceptualization of the isomorphies used in general systems science is dramatically unique and grants them a far more distinguished status. This newer version of the meaning of isomorphy results from the following critical questions. How is it that the same process/structure appears over and over again in natural systems, even though each system is clearly separated in time and space from the others, and exists on quite independent scalar levels of organization and complexity? Isn't it too much to ask of coincidence that it be responsible for this reoccurrence of the same form through 13 billion years of evolution and emergence? Why then should isomorphies recursively appear as each new level of complexity emerges in the long, concatenated history of systems origins?

A simple alternative conceptualization of isomorphy exists that would answer these questions: What if an isomorph was not just a comparison born of anthropomorphic searching? What if it was so fundamental a part of nature that it <u>preceded</u> "form" rather than being the abstract expression of similarity of form artificially derived after analysis of nature?

47

What if an isomorph was the most optimal configuration of interactions possible, the potential minima, the process/structure that required the least energy and time for stabilization? Thermodynamically, it would then be the most favored state. Whenever a very large number of parts on a newly emerged level of organization were experiencing random interactions during the process of systems origin (autopoiesis or autonomy-formation), they would inevitably and eventually settle on the same isomorphic interactions as the systems on other levels. Only later, much later, would man do the comparisons.

Figure 1 depicts this idea diagrammatically. An isomorph-specified interaction is like a potential well. The net of lines indicates the "field" of all "possible" interactions of the parts relative to a specific process for a new level of organization. Each "possible" interaction cohort is represented by one of the small spheres. The sphere at the bottom of the well represents the approximate isomorphic interaction cohort. Wherever, at whatever time and scale, and with whatever of the possible interaction cohorts a new system begins, it will suffer the same fate of stabilizing at the isomorphy. In terms of the diagram all the small spheres will eventually rest in the well. They will "migrate" until they become the interaction we later discover is common to all natural systems and so call an "isomorphy".

This suggests that isomorphies possess a far more fundamental, less anthropomorphic role, than their current usage allows even in systems science. Their current status as analogy and metaphor gives them a decidedly unreal dimension. Only the systems that are compared to find isomorphies are real in the conventional usage; the comparisons are mental; they are not as real as the systems. This reminds one of the age-old philosophical debate, begun with the Greeks, between nominalism and realism. Is the name (essence) of a thing the most real, or the physical thing itself? In the suggested new use of the term, isomorphies are real. In a sense they are more fundamentally real than their "manifestations" in so-called real systems because each new system at each new level of emergent complexity has to equilibrate through time before it stabilizes at the isomorphy. But the isomorphic potential exists independent of, and predates the physical manifestations. Undoubtedly there will be great resistance to this idea since it reverses our sense of what is real in science, and suggests a philosophical revolution. However, it is merely a macroscopic example of results that are appearing on the microscopic level of subatomic particle physics and which point in a similar direction.

Clearly, all the above is dependent on the rigorous empirical demonstration of isomorphies across all the levels of natural systems, a task which is still in its infancy. However, the number of workers adequate to accomplishing the task will not be attracted to the field unless the stakes are as high, and the rewards as significant as suggested here.

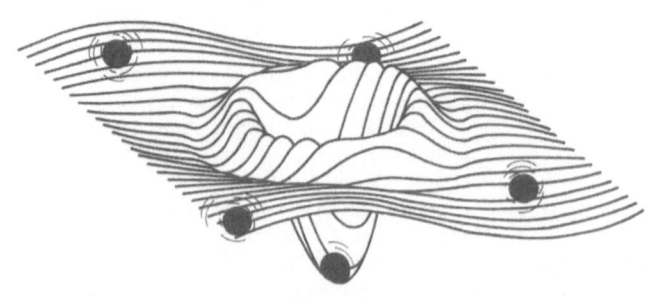

Fig. 1.

The purpose of the Linkage Proposition Model is the description of specific interactions among the currently identified systems isomorphies. This dimension of systems science has for some reason been quite neglected probably due to the incredible range of disciplines serving as sources for individual isomorphies (see set of 29 citations by discipline in Section 2.0). If the isomorphies are as real and as fundamental as described above, then the linkages among the isomorphies would also be an important area of study. To achieve this important purpose the first step would be selection of the isomorphies to be connected. This may seem too elemental a step to even discuss, but in this field it is a critical step. If at the onset of formulating the LPTM too many vague or non-phenomenological systems terms are included, then the important and "difficult-to-discern" isomorphic interactions that give rise to systems functions will be obscured beyond recognition. There will simply be too much "noise" to distinguish the "signal".

The progress of a science can often be measured by the growth and increasingly precise use of its terminology. Some say that the entire purpose of empirical testing is the more accurate definition of terms that humans use to describe their models of the world. Since general systems science is just beginning to use empirical approaches, its terminology is comparatively loose. No widely accepted criteria exist for inclusion or elimination of any particular term in its papers. At this early stage in the evolution of the field a multitude of terms abound, and even the same term is used in different ways. Both the nature of appropriate sources of the isomorphies, and the criteria determining what is and what is not a putative isomorphy are debatable. Worse, a significantly "conscious" debate on these fundamental issues does not yet even exist. Another purpose of the LPTM is to stimulate such a debate.

Although a definitive glossary of systems terms has not yet been published, several articles have attempted to list or analyze systems terms. Young described 36 systems concepts in 1964. Ackoff not only described 32 major systems concepts in 1971, but additionally cited the need for a "system of systems concepts" (Ackoff, 1971), which is an early recognition of the need for attention to linkages between systems concepts. Heinz von Foerster (von Foerster, 1974) defined 238 systems-related terms, mostly in the control theory sector of general systems science. The most weighty, and indiscriminate listing of systems (holistic)-related concepts is that of the Union of International Associations which defined 421 candidate terms (Union of International Associations, 1976). Recently, James Miller defined not only a multitude of systems-related terms in his thousand-page synthesis of information on living systems, but also called attention to "cross-level hypotheses", another recognition of the need to specify linkages between isomorphies (Miller, 1978). Rogers and Umpleby used the occasion of a computerized conference on General Systems Theory sponsored by the National Science Foundation to request participants to assemble an annotated glossary of systems terms. This has not been published to date. Klir and coworkers assembled a very valuable computerized bibliography of general systems literature which contains a keyword sort, as well as a permuted word-in-title sort, both capable of serving as a preliminary glossary without definitions (Gesyps, Klir, and Rogers, 1977). Though undefined, the words of this list have the advantage of being tied directly to their literature references. This work is being carried forward by the International Federation for Systems Research under the guidance of Trappl and coworkers (Trappl, Horn, and Klir, 1985). Jain sorted a sampling of systems concepts from the literature into 6 fundamental sets using Warfield's technique (Jain, 1981). Robbins and Oliva published a series of articles which statistically analyzed a sampling of the systems literature for systems terms and their usage. This resulted in a listing of 51 "key" systems concepts together with a clustering of these concepts by usage (Robbins and Oliva, 1982a, 1982b, 1984). This historical lineage of studies will certainly prove useful to the field in the future, however, none of them has been directly utilized in the selection of isomorphies included in the LPTM. The criteria for selection of terms to be used in the LPTM, as well as its distinct contrasts with the purposes and results of the above studies have been preliminarily discussed in earlier papers on the Linkage Model (Troncale, 1978, pages 31-34, 1982b, pages 28-30) to which we add the following.

If the LPTM is to model the real interactions observable in a wide range of natural systems, the isomorphies and linkages contained therein must include only those processes and structures having "correspondences" (Schaffner, 1969), or counterparts in real systems. Interactions can occur only between actual phenomena in the target systems. They cannot occur between inventions of the human mind, or symbols designed solely as aids to human thinking. Thus, all anthropomorphic, methodological, and taxonomic terms of the field must be rigorously excluded, although their presence is quite appropriate in the above-cited glossaries or samples of the working literature. Simply stated, a phenomenological model must exclude human-based jargon because it is not real; it does not have independent existence. The one important exception is the modern recognition that man is part of all of his models because he builds them. That insight is built into the model in its linkage propositions concerning autopoiesis, self-referencing mechanisms, recursion, and resonances (and their interactions).

Only the minimal set of isomorphies should be utilized in respect for Ockhams razor. There must be some attempt to provide proof that the "full" minimal set is used. Isomorphies that have been shown to be true for only two disciplines, or a limited scale of organizational complexity should be excluded pending future indications of their transdisciplinarity. Isomorphies included must also have the appropriate level of abstraction such that their expression admits co-mapping to a broad range of the unique and different expressions of particular systems. According to these last two criteria, for example, some of the concepts and cross-level hypotheses of Miller would not be admissible because they are restricted to living systems alone (Miller, 1978).

Every isomorphy included in the LPTM should be the name for a recognizable process in nature. Process, not product, drives the universe (Troncale and Wilson, 1977). Despite man's penchant for goals, objectives, and purposes, research into all natural systems except for man and his engineered systems indicates that goal/purpose does not exist in nature. Man continues to project his conscious goal-orientation on nature in a host of "isms" (creationism, vitalism, teleology, in some ways even holism), but much more economical models of how things appeared are sufficient. The century of debate and experiment surrounding evolution and analysis of the biology and ecology of all creatures up to man supports the thesis that, in nature, "process" is the driving force. The beautifully efficient and complex organization of social systems has been a fertile field for those who would explain the order in systems by the short cut of using purpose. But due to the above considerations and others (Troncale, 1985a), "purpose" as a cause, should be eliminated from general definitions of systems , including social systems, in favor of the more neutral term "function". Function requires the identification and inclusion of forces from the context/environment of the system studied, rather than the blurring of these real forces behind the phantom of purpose (Troncale and Wilson, 1977).

Isomorphies used for interconnections in the LPTM should also be "anasynthetic". This is an invented word which draws attention to the observation that isomorphies and their linkages as used in the LPTM are analytical and synthetic simultaneously. They break the normally vague concept of wholes into many highly detailed sub-interactions (analysis/reductionism) while they simplify the potentially infinite number of real systems in nature down to a single model of limited interactions (synthesis). A related criterion for selection of isomorphies judges whether or not an isomorphy is useful in simplifying complexity as a measure of the computational and informational explosion that occurs as one progresses up the ladder of systems evolution.

Two other criteria used to select terms from the systems field for inclusion in the LPTM are the qualities of self-definition and robust interaction. Self-definition means that any isomorphy selected should generate a list of linkage propositions that significantly helps define the isomorphies already selected. That is, the mechanisms that are elucidated by the linkage propositions which are required or enabled by the added isomorphism should partially result in the appearance of the other isomorphies. The set defines and explains itself. Robust means that any selected isomorphy should demonstrate numerous and specific couplings or influences with already established isomorphies. Robust, as in mathematics, means that much more is produced than was added.

Both of these last criteria are phantoms. To judge a putative isomorphy in either case requires a highly conditional assessment, because the existence of the full set of isomorphies and an elaborate set of linkage propositions could markedly influence the judgement. The existence of such a situation greatly infuriates reductionist scientists who use it to attack the rigor of this field. I'm reminded of Bohr's tactic of insisting at conferences that when a photon approaches two slits, it goes through both, until this impossible thought was finally accepted. Is particle physics, therefore, a non-rigorous science? The above Catch-22 type of paradox is typical of systems studies. I call the necessary technique that results from these paradoxes, "mutual selection". That which you are studying effects the context which effects that which you are studying. The only possible solution is a recycling set of tests and judgements that results in a gradual, recursive evolution and optimization. And this is why I use the term "empirical refinement" rather than the term of empirical verifiability/falsifiability. As infuriating as it is to conventional reductionist scientists, and as frustrating as it is to systems scientists, this feature of systems science is still as real and necessary as it is in the natural systems which systems science studies; it needs to become acceptable and accepted feature of science in the 21st century.

It is hoped that this explicit naming of a dozen criteria for the all-important step of selecting which systems concepts should be examined for linkages will lead to a conscious debate of both the criteria and the selections. Table 2 shows a current listing of about seventy candidate isomorphies that survive preliminary application of these selection criteria.

Table 3 is a listing and definition of some of the fields surveyed to obtain the listing of isomorphies. The point is that all of the fields are appropriate sources of putative isomorphies. Some isomorphies are more observable on one scale of reality than others (given man as the observer), while some linkages between isomorphies are more tractable or measurable in still other fields. Theoretically, the LPTM should be equally applicable to all fields; they all study systems of one sort or another. Clearly, the depth of application will be markedly different for each.

Surprisingly, the more holistic-type intellectual movements from cybernetic theory to simulation to information theory and operations research are not currently the most fertile shopping places for the discovery of isomorphies. Currently the most productive fields for recognition of isomorphies are the physical sciences. However, the potentially most influenced fields are the holistic intellectual movements. The most important insight achieved from this mapping comes from its use as one axis for still another study that has, as the other axis, the seventy putative isomorphies. Many fields have no recognition whatsoever of certain isomorphies, but may have elaborate expositions of others. For example, one such study conducted at our Institute as part of a much larger project indicated that the field of conventional economics did not mention hierarchical structure in any of its many textbooks (Troncale et al., 1976). Consequently, it was not until recently that clustering of operating units appeared in economic models, or world game models. Many other such "gaps" of recognition of isomorphies, their consequences and applications appear when disciplines are compared. The matrix of disciplines versus recognized isomorphies (if ever completed) would be a dramatic "periodic table" whose unfilled interstices would expose the potential future developments for any field. Young workers, using the matrix of such comparisons, could significantly contribute to a field by introducing a formerly ignored isomorphy to the field, and capitalizing on its many interactions (already suggested or specified in the LPTM) with phenomena of the field. This application is an offshoot of the overview work typical of general systems science which has the responsibility of collecting and comparing knowledge across a span of literature witnessed by no other discipline.

2.3 Towards A Taxonomy of Isomorphies

It could be argued that general systems science is at about the same stage as the biological sciences were before Carl von Linne. Many organisms had been described at that period, but there was only a vague notion of their interrelationships compared to what later developed. And most importantly, all the dynamics of the ecosystem were absent. In fact, the first taxonomies of Linnaeus were constructed of unchanging, immutable, eternally constant organisms designed for their place in nature by a purposeful creator. Evolution by natural selection during those times meant that any organism deviating from the plan was eliminated. Natural selection actually accounted for the lack of dynamics in the system.

Ironically, it was this same, completely static taxonomy that enabled later workers to conceptualize a dynamic, change-oriented evolution. Without the initial ordering, without the immense amount of attention to details of each organisms' environment, without the painstaking studies of comparative anatomy, and without the improvements in organization of the information, the forerunners to Darwin would not have possessed the intermediates necessary - the catalogue of gradual changes necessary - to recognize the possibility of evolution.

Taxonomy preceded dynamics. Taxonomy preceded conceptualization of a synthesizing theory. And where are we in general systems science? No adequate taxonomies exist at all. The five attempts that I am aware of only "nibble" at the problem from one perspective or another. Klir and coworkers have a taxonomy built into their general systems problem solver (Klir, 1984) and current efforts in Reconstructability Theory (Hai and Klir, 1984). Oren has attempted a rather detailed taxonomy of the limited domain of systems models and tools (Oren, 1985). Miller has a partial taxonomy in his work on living systems although it is very general compared to the others and leaves out most natural systems (Miller, 1978). Bunge has taken great pains to develop a philosophically systems-based taxonomy in his work (Bunge, 1959, 1972). And to this we may now add the taxonomic "nibbling" inherent in Table 2. There is a very great need for integration of these various attempts, accompanied by detailed debates across the teams aimed at further clarification of performance criteria and techniques. The absence of a consensus taxonomy is a major obstacle to further development of the field (Troncale, 1985a). Table 2 is premature as a taxonomy. But it may illustrate the potential. Isomorphies could be clustered into taxonomic categories according to their contribution to realization of fundamental systems behaviors or functions. In this approach, the isomorphies would be clustered by direction of their linkage propositions. This is a phenomenologically-based taxonomy traceable to the processes in actual systems, not a human-usage-based taxonomy as in the studies of Jain (1981) and Robbins and Oliva (1982a, 1982b, 1984).

It is also a taxonomy with significant "dynamic" and "functional" meaning. The isomorphies listed together in one cluster presumably couple more tightly than those listed in another. Their mutual influences within the cluster and between the clusters are specifically stated and, thus, are more tractable in the LPTM. Although they may play some secondary role in assisting stabilization of other systems functions, their primary intra-cluster influence is in conjunction with those of their category in determining a recognizable and critical systems behavior. This dimension of specifiable dynamic function allows initial description of the "limits" of each isomorphy. Describing the limits of what we know may be more important to the maturation of a science than describing what we think we do know. Most importantly, the measure of this taxonomy would be its ability to aid in explanation, understanding, and utilization of fundamental systems functions as its primary features.

Table 2 is a hierarchical listing as are most taxonomies. It is designated premature as a taxonomy because this aspect of the LPTM has not been studied directly and extensively. Consequently, the depth of the version shown is only 3 or 4 levels, hardly an instructive taxonomy (about seven levels of depth exist but are not shown in this version). Still several observations are already possible.

First, systems science may not be able to achieve, or even delude itself into thinking that it may achieve, a single, correct taxonomy. It is generally assumed that in a hierarchical taxonomy an element has one, and only one correct placement. This may never be true of systems science and illustrates some of its differences from regular science as well as its potential contributions to the changing conceptualization of science. Some of the isomorphies have impacts on several general systems functions and so could be equally correctly placed in more than one category. In fact, they must be. For example, the principle of plenitude plays important roles in systems form/structure as well as systems growth and developmental processes. All of the types of cyclical behavior could equally well be placed under systems flow processes as well as systems transformation processes. Solitons could be placed under cyclical behavior except for someones intuition that they are telling us something specific about field characteristics. In fact, it is exactly these types of deep interdependencies that give rise to the linkage proposition model. Even the higher order systems functions themselves may be rearranged or changed. Perhaps long-term work with an adequate number of linkage propositions will allow best case placements, but the fundamental point remains that this is a chameleon taxonomy itself capable of surprizing transformations.

Second, the work of mathematicians and programmers indicates that hierarchical structures and network structures are co-mappable and co-transformable. If one selects out a node of a network and makes it the origin point, one can transform the net into a hierarchical tree, and vice versa. This is especially true of this taxonomy of isomorphies. It is a hierarchy that with the linkage propositions in the LPTM creates the nets we are about to examine in the next section. An important related question comes from studies in computer science of the somewhat conflicting optimalities of nets and trees. Each has advantages the other has as disadvantages (perhaps isomorphically reminiscent of the complementary advantages and disadvantages of protein and DNA which initiated the biosphere on planet Earth) (Troncale, 1985b). Is it significant that the core of general systems structures and processes also has this complementary feature? Undoubtedly, this is the mechanistic forerunner of systems origins, hypercycles, autopoiesis, and autonomy.

Third, changes to this initial taxonomy are necessary. For example, category 1.0 should be eliminated and its rather weak candidate isomorphies allocated to the various other categories. This cluster has previously served as a catch-all category and does not map very well with the others in that it is the most anthropomorphic-based and non-phenomenological function listed. Ironically, it contains some of the concepts most often used as the beginning of other taxonomies.

Fourth, rearranging the <u>sequence</u> of categories shown in Table 2 suggests the possibility of specifying, for the first time, a primitive life cycle for systems in general. Historically it can be shown that man comes to recognize the life cycle of any real system in a discrete series of steps. First he begins with a very slow growth in recognition of the phenomena, followed by a gradual accumulation of typical examples of the phenomena, then follows a static categorization of the types and intensive research into each of them. Only then, and only slowly, do dynamic interchanges become recognizable and these finally result in a rearrangement of the types into one flow that culminates in recognition of a "life cycle". This has been the order of discovery for recognizing the life cycles of organisms, the life cycles of cells, the life cycles of stars, and now possibly even of galaxies. Social scientists are studying possible life cycle stages of the individual psyche, families, corporations, and given Toynbee, even of civilizations. Systems science is just now beginning the stage of static categorization of types. With the mechanistic elaborations provided by the accompanying linkage propositions, the sequence of life cycle stages suggested by Table 2 achieves a new level of specification and utility. Figure 2 shows the putative stages in the generalized system's life cycle which would be: constraint fields and potential fields of system field characteristics, systems origin, appearance of form, stabilization and maintenance of form, superelaboration of linkages, reorganization of linkages into internal and external flow processes, establishment of growth and development patterns, projection of meta-level systems field characteristics, and finally, systems decay processes. Figure 2 shows not only these stages but two other elements of the generalized systems life cycle. Just as in the case of stellar life cycles (wherein main sequence stars may experience three alternative end states depending on initial conditions and field characteristics, these being supernovae, neutron stars that become dead bits of matter, or black holes) so also systems may end in decay, evolution to altered systems that survive in different form, or emergence onto an entirely new plane of organizational complexity, there to begin the cycle anew.

At this point it would be fascinating to explain how one might trace in detail the mechanics by which the isomorphies in any one taxon result in their designated systems function. This could be attempted using the linkage propositions, and the literature results from the various disciplines listed in Table 3. However, any single such case study would require an entire chapter of a book and is beyond the scope of this article, though such a book is in progress (Troncale, 1981b).

3.0 THE LINKAGE PROPOSITION TEMPLATE MODEL

The difficulties in establishing a single, supportable taxonomy for the isomorphies leads us directly to the importance of the mutual and multiple influences each isomorphy has on the others. It is exactly this dimension which has not been explored by general systems theorists to date, but which has the potential of becoming the highly detailed and useful "special knowledge" expected of established fields. The problem has been one of technique. How could one make the many interconnections between isomorphies practical and concrete? How could one render them capable of manipulation? How could one keep track of them and their "network causation?" How could one make them specific and discrete enough to allow tracing of their influences in a way that would free holistic studies from their brand of being "flaky" and vague, yet still keep them truly integrative? Endless prose narratives describing the same small set of isomorphies and only occasionally their interconnections have been neither adequate nor tractable.

3.1 What Is A Linkage Proposition?

Recently, the concept of "Linkage Proposition" was introduced (Troncale, 1978). A Linkage Proposition is a semantic statement of a single, specific interconnection or mutual influence between two or more isomorphies. The criteria for a Linkage Proposition are simple. It is a

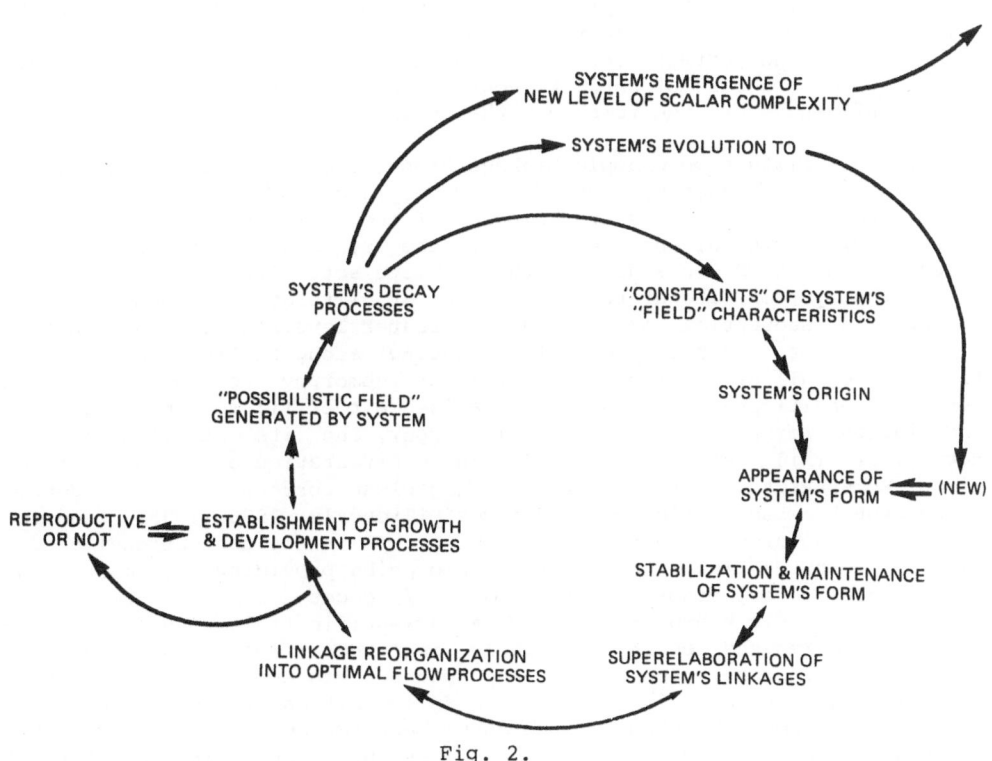

Fig. 2.

highly abstract, initially qualitative statement on a theoretical level which tentatively holds for all known observational entities in real systems that correspond to the theoretical construct. Stress on the singularity of the Linkage Proposition (hereafter LP) insures later tractability and testability. However, it is also important to stress that the logical statements typical of LPs are not necessarily causal, and are certainly not linear, nor directional as is assumed to be the case for statements in normal science. The names of isomorphies must be included in the statement (usually signified by underlining) and each isomorphy must be separated from the other by a discrete phrase which specifies the interconnection or influence.

The LPs are termed "Linkage" Propositions because they tie many isomorphies together, because they mutually generate each other (are self-organizing), and because the entire set serves to define itself (self-defining and self-limiting). The result of all of these linkages is a complex network graph of systemness as seen in Figure 4, where the nodes or small spheres are isomorphies, the large spheres are major systems functions (not usually isomorphies, but rather their concerted result), and the various categories of lines represent Linkage Propositions. The result is an initial version of the much-needed "system" of systems concepts.

The LPs's are called "Propositions" because in many cases they are unproven hypotheses on the order of mathematical conjectures, or intuitive relations derived from the literature of the natural or systems sciences. LPs are considered unproven even when supported by the experimental evidence of a conventional hard science discipline because the evidence is available from only one discipline. Due to their usage in a general systems model, the interconnection described in a Linkage Proposition must be transdisciplinary, which is to say experimental evidence must be available in a series of disciplines before an LP may be considered "empirically refinable". They are also propositions because they could not even be tested by the normal experimental method. This requires isolation of a linear cause and effect set from all other influences, which is clearly not possible in a case of network, non-linear causality where numerous and mutual influences are the rule, not the exception.

Table 4 lists some example Linkage Propositions. As explained in a previous paper (Troncale, 1978, 1982b), the large number of LPs currently under study (n = about 200) are composed of the following types. Some are already well known in, and supported by the systems literature. Their value in the LPTM is adding to the complete set. Others are completely new to the systems literature, or any other literature. Some can be arranged in sequential order adding greater levels of explanation to a systems concept. This can be called "tracing" along an "influence vector" through the full set or from one particular isomorphy to another. Some describe conditions that must be realized either (i) for an appropriate and rigorous formulation of another isomorphy, and its stabilization in nature, or (ii) as a condition for the manifestation of a systems function. Some LPs are useful for recognizing close correspondence or identity between two isomorphies or "discinyms" (Troncale, 1976, 1985) used by relatively isolated, non-communicating groups in the conventional disciplines. Some LPs link more than two isomorphies providing for multi-concept "traces", or cohort actions and influences. Some LPs are derived from a result well known in a particular scientific field generalized to a level that makes the result applicable in other fields in which the relation is currently completely unsuspected. In any case, the examples of Table 4 are included in here only as an indication of the nature and potential of the LPTM. None of them have been refined or subjected to the scrutiny of many clever competing minds as they must be in the future if the LPTM is to evolve into a usable tool.

A final note about LPs. They are fully as "isomorphic", by definition, as the isomorphies they link. Consider Figure 3 as an extension of the idea presented in Figure 1. Again the network of lines indicates the potential field of all possible interaction configurations available between two isomorphies. The two large spheres in the wells are isomorphies as explained in Figure 1. The bar within its own well represents an LP-specified interaction between the isomorphies. As in Figure 1, any initial condition of a new system with a multitude of parts will begin spontaneously producing interactions among its parts (represented by the small connected spheres on the field lines) and these interactions will seek optimal configurations. In terms of the diagram this means they will eventually slip into the thermodynamic well occupied by the bar which is the graphic representation of one of the Linkage Propositions between the two specified isomorphies. This scenario applies of course only to symbolic representations of LPs that have been refined by decades of research.

3.2 Association Classes of Linkages Propositions

The total number and limits of LPs is as unknowable at this early stage of study as the total number and limits of isomorphies. At present about 70 isomorphies and nearly 200 Linkage Propositions are under examination. This level of detail should be welcomed by a field often attacked for lack of substantive results. Oddly enough, it is the LPTM that has been attacked from within the systems movement by some who consider its formidable detail as anti-holistic, anti-synthetic, and creating complexity where it does not exist. Apparently, some workers associate holism with something as all encompassing and simple as a "mandala" or "om-word". Clearly, the numbers of elements of the model cited above indicate that the mature model will be even more detailed and thus more complicated. Yet as complex as it is, it is still immensely more simple than the legion of parts and interactions of real systems from the microscopic to the macroscopic scales of the universe - all of which are covered by aspects of the one model. In this lies its integrative, synthetic, holistic, and simplifying power. It is simply not true that holism must be vague, or else be accused of being reductionist.

Part of the complexity results from the immature stage of development of the model. For example, the first 140 LPs were examined for similarity of phrases used to describe the linkages and about 15 phrases were found to be used over and over again to link different isomorphies (Troncale, 1978, 1982b). These could be further grouped into just four major "association classes" of Linkage Proposition (Troncale, 1982b). As the model matures, these simplifying techniques will also mature rendering the

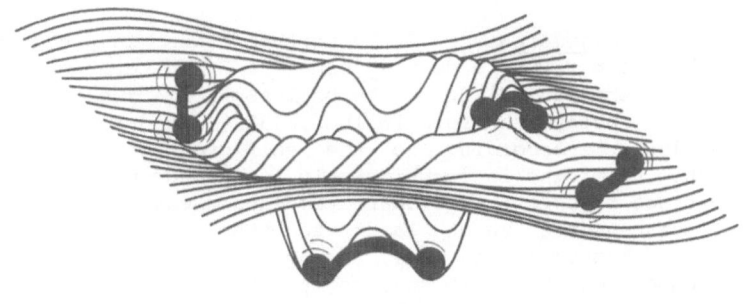

Fig. 3.

LPTM more manageable and utilitarian. It should be noted, however, that the discovery, significance, and utility of these association classes and key phrases (especially those presumably new to science) could not have occurred without first going through the more detailed stages of formulating, testing, examining and clustering the initial Linkage Propositions. Further, the recognition of these classes are critical to the development of relational data bases of the LPs, mathematical formalization of the LP set, and its use in expert systems (see Section 5.0).

4.0 PROJECTED USES AND PROJECTED CRITICISMS OF THE LINKAGE PROPOSITION MODEL

The time, energy, and manpower required to develop this model demands an early assessment of whether or not its use justifies its costs. Such an assessment should also try to describe anticipated criticisms for the same purpose. If the criticisms seem insurmountable, the effort may be unjustifiable. Equally important, early criticisms can often be used to improve a research program, during its development when some of the most direct and quick responses are possible.

4.1 Uses of the LPTM

Section Five cites five detailed case studies of anticipated uses for the Linkage Proposition Template Model (LPTM). They are not included here, and for that reason this begins at the sixth use. All projected uses assume the existence of a mature LPTM.

(6) Definition of Isomorphies - the LPTM would enrich the definition, meaning, and understanding of each isomorph included in it. The dozens of LPs designating the particular mechanisms by which a given isomorph effects others, or how it is effected by others, provides a richer understanding of the function of that isomorph in the total set. The role of a given isomorph in producing a major systems function/behavior or part of the general systems lifecycle are also explained in greater detail by the LPs. The set of LPs coursing to and from a given isomorph also "actualize" formerly vague concepts. Let me exemplify this using the concept of "entitation". Gerard, a neuroscientist and founder of the Society for General Systems Research, used the word "entitation" to describe the expanded definition of an entity that included all of its phase shifts, developmental changes in time, and all of its connections with other entities during its lifecycle. He was fond of criticizing reductionists (though a practicing reductionist himself) for their assumption that they could empirically capture an entity with physical measurements only and at one given moment of time. Clearly, all of the LPs attached to a given isomorph (see Figure 5) are a detailed manifestation of the entitation of that isomorph, and therefore a more complete definition in Gerard's terms.

(7) Origins of Isomorphies - If isomorphies are as fundamental as posited in this paper, where do they come from? In terms of the natural sciences they are the equivalent of the philosophical "uncaused" causes, or the emptiness before the form (Wilber, 1982, Chung-Yuan, 1969, Capra, 1975, Merton, 1965). According to the self-organizing, self-defining feature of the LPTM, the isomorphies only appear uncaused because of our inability to trace or specify a linear, time-dependent, stage by stage, cause-and-effect sequence which leads to them. Rather the entire set moves "gradually through a great age, trying all manner of combinations, until those come together that stay together" (in the prophetic words of the Roman philosopher Lucretius in De Rerum Novarum). This is not an uncommon occurrence because we find it even in the phenomena in conventional

science, much less systems science. Witness the new results on the community of quarks assembling primitive matter or the new results on the initiation of genetic processes from primitive RNA and protein polymers in protobionts, or the induction of hypercycles at the origins of biochemical pathways. We should expect this self-organizing feature to be true of the general model if it is so common in the physical manifestations of the general model we see in our disciplines. And the existence of the LP set specifies the network of mutual influences that accomplishes this stabilization of complexity, although it expresses them on the abstract, transdisciplinary level.

(8) <u>Dynamic Process-Orientation</u> - Man is enamored of stability to such an extent that he subconsciously worships static states. He is cautious of change whether it be in ideas, conditions, or behavior. As a result, the history of science, and of civilization as well, is often measured by painful progress toward reluctant recognition of the predominance of the very dynamic process-orientation of nature (Troncale and Wilson, 1977). Static taxonomies of organisms become evolving ones, astronomical entities fixed in the sky become perilously loose; we leave our families, loved ones leave us. In the same manner systems science, even though based from the start on dynamics, has erred in too little emphasis on or too vague a specification of the full extent of the mechanics of systems interactions. The swarm of connections shown in Figure 4 coursing to and from each isomorphy should be another step in the direction of establishing the rightful place of dynamic process, not fixed purpose or product, as the hallmark of all natural systems.

(9) <u>Enhance Search for Isomorphies</u> - The existence of a set of isomorphies that define each other could reveal the absence of potential isomorphies, gaps in or inadequacies of a required self-definition; a sub-set of isomorphies poorly connected by LPs to the full set; or inability to explain a critical systems function through its isomorphies would signal possible ignorance of a potential isomorphy. The existence of the many specified LPs could be used as an aide in predicting, or visualizing absent isomorphies in a manner at least qualitatively similar to the prediction of elements or sub-atomic particles.

(10) <u>Empirical Refinement</u> - The detailed mechanics and influences embodied in the LPs should improve chances for design of concatenated tests across a range of disciplines, each in the manner acceptable on its scale of reality, leading to better resolution and refinement of both isomorphies and LPs. Notice that there is no ambition here for verification or falsification (Popper, 1959). The existence of mutual influences and self-organization, as well as the participation of the observer in the system precludes such ambition. Nevertheless, much can be accomplished toward improving the utility of the LPTM by attempts at empirical refinement.

(11) <u>Generalized Predictions</u> - The above mentioned constraints on empirical testing nullify use of the models of general systems science for the types of predictions, and calculation of limits on predictions typical of the physical sciences. Still, tracing of multiple influences can lead to an understanding of classes of consequences. The existence of detailed mechanics can lead to a heightened awareness of key measures or signals to watch for, or the extent of resilience to disturbance typical of certain classes of systems. Very general features of descendent systems or emergent systems could be predicted given increased understanding of the relative impacts of isomorphies and LPs.

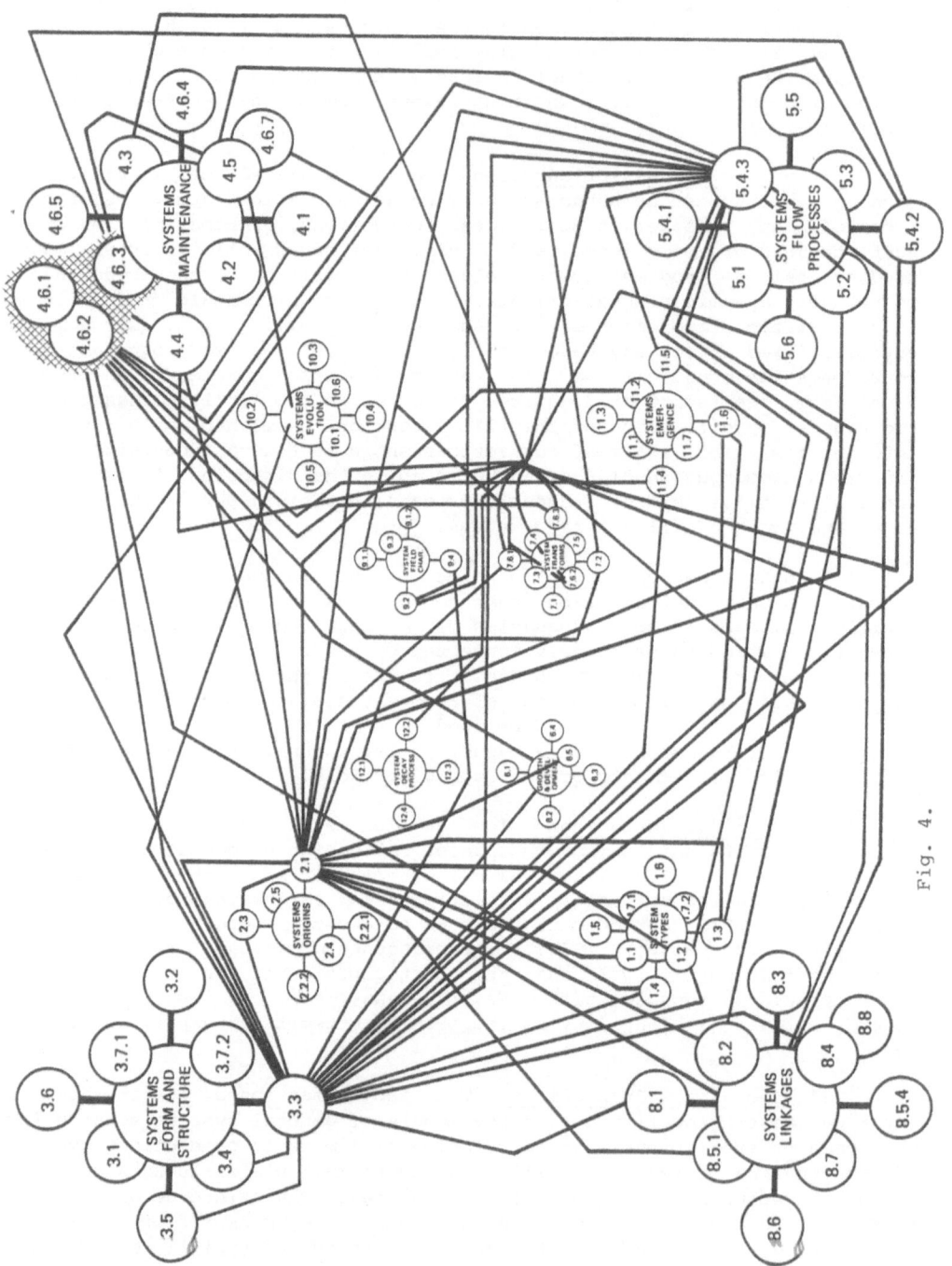

Fig. 4.

60

(12) Research Management Tool - The detail of the LPTM could help investigators in the field of systems science determine on what aspects of the field they are focusing, for example on which isomorphies, or LPs, or combinations. It could help them identify who are significant colleagues and collaborators in those areas. It could be a very significant organizer of not only the systems literature, but also pertinent results from the literature of the conventional disciplines. If widely shared, it could be the framework of a group paradigm guarding against duplication of effort or loss of results in the mass of results.

(13) Detection of Areas of Neglect - Once a sufficiently detailed image or morphology of a class of objects is known, it can be used to detect missing knowledge. The deficient areas can be the subject of increased interest and investment of time by new workers seeking service, or a reputation, or they could be the target of increased funding. For example, poorly known stages of the generalized systems life-cycle (Figure 2) could be targeted for increased efforts [see the fascinating uses of General Morphology in this respect for examples from the physical sciences. (Zwicky, 1969, Zwicky and Wilson, 1967)].

(14) Checklist Evaluation of Other Models and Simulations - Many disciplines now have a small cadre of workers who are constructing systems analytical simulations or systems theoretical models of phenomena in their domain (see Table 3). Biology, for example, has a significant literature on systems analytical simulations or models of virtually every level of bioorganization from molecules to cancer cells, to development of organisms, to lake ecosystems. In addition, its established professional societies (such as the Ecological Society of America) have spun-off new societies devoted entirely to simulation, e.g. I.S.E.M., the International Society for Ecological Modelling. Systems analysis is big business even in fields other than computer and information sciences. But many of the models constructed are biased by the reductionist environment in which they arise. Most are designed "bottom-up"; their model components are included as they are encountered in reductionist experimentation. This normally results in important and necessary sectors of a model being left out since they simply have not yet been elucidated by the limited scope and steady grind of reductionist approaches. The LPTM will provide for a modest "top-down" approach to serve as a necessary counterbalance (Troncale, 1985a). With its many isomorphies and LPs it could serve anyone doing systems analysis by providing a holistically more comprehensive and complete checklist to compare against. Phenomenologically-based models in the disciplines should have counterparts for most of the elements in the presumably more complete general model of the LPTM, otherwise they are incomplete. Undoubtedly, this feature will irritate real systems modelers who often delude themselves into thinking that their discipline knows all there is to know about their system already. This indicates that the LPTM may be even more important to decision-makers who must evaluate these incomplete, disciplinary-based models for use in the real world.

(15) Efficient Education of Students - With some of the computer assists suggested in the next section, the LPTM would be a very utilitarian teaching tool for any of a number of systems science, and general systems science programs, or for systems analytical portions of many discipline-based curricula. Even in written form (Troncale, 1981b), it would be useful as a text for such curricula. The advantage of the computerized version is its adaptability for self-instruction and student research. No competitor offers as many isomorphies with as many interconnections between them. A possible disadvantage might be its level of complexity, although parts of the model were originally utilized for the design of a teacher education program in applying systems concepts to environmental

education, K through 8th grade (Troncale et.al., 1976).

(16) Built-in Rules for Deabstraction - A little-recognized obstacle to progress in general systems science is the absence of a technique for moving in the opposite direction from that used in devising the models. Most do not include unambiguous "rules for deabstraction" or "scale translation protocols" (Troncale, 1985a). Yet these are a critical necessity. It is very difficult to apply rules that had to be formulated at such a very general level of abstraction that they achieved sufficient transdisciplinarity. The many LPs between isomorphies will have little descriptive, diagnostic, or prescriptive power if the counterparts of the isomorphies cannot be recognized in the application domains. The coupling of the LPTM research projects to another project series at our Institute involved in collecting and organizing data bases from the conventional disciplines (Troncale, 1981a, 1982c), might help make the LPTM more amenable to deabstraction. Much more than this, however, will be necessary.

(17) Toolbox and Toolmaker - As recognition of isomorphies continues to grow, the argument that virtually all current systems analytical tools gain their power from their incorporation or capitalization on one or a few isomorphies may also grow in acceptance (Troncale, 1985a). At this point, general systems science will be seen as the toolbox for systems analysts (but only if the field progresses rapidly beyond its present state). The many isomorphies and LPs of the LPTM at that time would be a useful tool for further toolmaking and refining.

The existence of the above 17 different uses of the mature LPTM implies that time, manpower, and moneys expended in its development might not be wasted.

4.2 Disadvantages and Criticisms of the LPTM

The shorter length of this list is not to be regarded as proof of the value of the LPTM. Future criticism will surely lengthen the list. The model will be improved only if as much energy is expended in trying to find out what is wrong with it, as is put into developing it.

(1) LPs Are Unproven. The most damaging criticism is the hypothetical state of the Linkage Propositions. Even if an LP has been proven true in one discipline, or several, it must be regarded as unproven for the remainder until we have some way to provide evidence for true transdisciplinarity. Further, this proof can never be absolute, or of as high a certainty as we have come to expect from some sciences. Can such unproven models be useful? I believe the answer is yes, if it is used cautiously as a checklist, or as a design tool, or to stimulate creative ideas, or to educate novices, or to diagnose various systems problems. Actually at present all models, even those in the most empirical sciences, are unproven and have only "relative" uses albeit still powerful and productive ones. The danger may be that the detail inherent in the model might lull those in need or those unaware of the limitations of models into a prescriptive use beyond its capability.

(2) The LPTM is Too Complex for Humans To Use. Section Five, in its entirety, is addressed at solving this problem.

(3) Short-Term Human Systems Are Immature. The LPTM is a model of mature natural systems, systems that required many millions of years to equilibrate, systems that possess the full set of isomorphies, systems that would disappear unless their mechanisms of transformation, evolution, and emergence were not inherent. Human-based systems are simply not this long-term in their duration or their intent. They do not possess these

features. Perhaps we all would be better off if they did. That is the reason for the title of this paper, and the message of this book. On the other hand, it would be ridiculous to spend the amount of time and money necessary to design a car that evolves. (Can you even imagine one that emerges to new levels of complexity? Actually, we all may wish we had one to replace the one in our garages!) Man thinks he needs cheap, short-term systems that answer his material needs; these do not require the LPTM. But still the LPTM could be used in these cases for ideas and some fundamental guidelines; it just could not be directly applied as a template.

(4) The LPTM Is A Bootstrapping System. Man has had great difficulty understanding and applying "bootstrapping" systems even when they have been encountered in the most rigorous of the physical sciences, for example, subatomic particle physics. Perhaps this is not a crippling objection since the LPTM might help achieve understanding of this important, pervasive phenomenon by showing in specific ways how "bootstrapping" systems are initiated, stabilized, and maintained.

5.0 FUTURE WORK ON THE LINKAGE PROPOSITION TEMPLATE MODEL

Some of the following research programs are underway, although with very little funding. Others have progressed only as far as a feasibility study. Collaborators are needed.

5.1 Graphic Representation of the LPTM

Because of the great complexity of detail inherent in even the currently primitive version of the LPTM, it was decided to present the model as a network graph. Figure 4 shows what the LPTM might look like in graphic form, showing LPs connecting only five isomorphies. The large spheres are the major systems functions. It is these that manifest the survival of systems such that man can observe them. However, these systems functions are not described as isomorphies, because they are all-encompassing, global terms dependent upon more specific mechanisms. These more specific mechanisms or processes are called isomorphies. They are shown as the smaller spheres associated with each function. The Linkage Propositions are shown as the many rods connecting isomorphies and functions together. Different types of rods could be used to reflect the different association classes of Linkage Proposition. Not all LPs are depicted here to simplify the overview, but even at this simplified level the detailed complexity may inhibit comprehension.

The biology of humankind is especially well-adapted for utilization of pictorial representations such as this. The popularity and success of the Macintosh computer is based on this attribute. Man's survival in a topological world is another example. The description of the LPTM in all of its sordid detail would certainly take a book the length of a dictionary, and might be just about as interesting. Anyone wishing to learn it, or use it as reference would easily get lost. But Figure 4 could contain as much detail, be interesting and pleasing to explore, and reduce significantly the danger of loss of orientation with the resulting confusion.

One technique used to simplify presentation of the graphic LPTM is the technique developed by systems programmers and systems analysts called "stepwise refinement" and "graduated entry". In these techniques users are presented with only the most general interrelationships first, and then step-by-step with more detailed versions. The LPTM would be viewed first only at the level of interrelationships of the major systems functions, or the stages in the generalized systems life cycle. Then a user might choose to view but one function and its interconnected isomorphies ignoring the rest for the moment. Gradually, as the user's understanding and orientation becomes more enriched, additional isomorphies with their interactions are added until the full model is explicated. This is also of obvious value as an heuristic learning program. The upcoming text that explains all of the isomorphies and the LPTM will use this technique throughout as its graphic component (Troncale, 1981b).

5.1.1 Isomorphies are Real (Part II)

This same diagram can be used to illustrate an extension of the argument that isomorphies are real (begun in section 2.1). Recall that Linkage Propositions are purported to be as isomorphic as the isomorphies they mutually influence. Therefore, the LPTM is hypothesized to represent the most conservative interconnection of any multitude of parts and interactions vis a vis their use of space, time, and energy. This cannot be easily shown in a diagram. To show that two isomorphies connected by a Linkage Proposition are a highly likely probability state we depicted them as a well in a thermodynamic, potential field. To do this we had to use a three dimensional diagram of the simple two dimensional ball and stick picture (Figure 3). Now imagine the entire three dimensional LPTM as shown in Figure 4 as the most probable thermodynamic state. To see this you must visualize an entire complex network of depressions or wells for each LP and isomorph in a deformable, pervasive potential field of a higher dimension. Man cannot visualize these higher dimensions directly; he can only perceive their "shadows" in terms of their intersection with our three dimensional universe as has been demonstrated for a limited number of cases in recent AI research. If we could see such a diagram, it would visualize how the entire LPTM is favored as a system reaches "maturity" whatever its original beginning state.

5.2 The Graphic LPTM and Computerized Data Bases

Planned computerization of the graphics described above could yield very powerful benefits. Levels of "stepwise refinement" could then be specified and received in real time by any user according to their special needs and instructions. The computer would allow multiple entry points catering to the users particular interests. Various levels of specificity could be flipped back and forth for a dynamic real-time visualization of change yielding a "developmental" impression. The computer would have a symbolic representation of the graphic LPTM built-in which would allow conversion of any "sphere" representing an isomorphy, or "bar" representing a Linkage Proposition, into its corresponding natural language statement. It would also possess a graphics generator compatible with this translator. With these a user could produce a number of personally tailored diagrams by issuing commands representing the following requests:

* Show me all of the Linkage Propositions connected to the isomorphy, "Boundary Condition." (see Figure 5) This could be requested for any isomorphy of choice.

* Show me the linkage net between the following two isomorphies (or any specified, small combination of isomorphies).

* Show me the LPTM with only the Linkage Propositions of one specified "association class" of Linkages included. Or one "type" of LP. Or one LP function.

* I am at one isomorphy and want to follow a "trail" of Linkage Propositions wherever it leads from one isomorphy to the next. (This could be followed in an interactive series of displays guided by the computer).

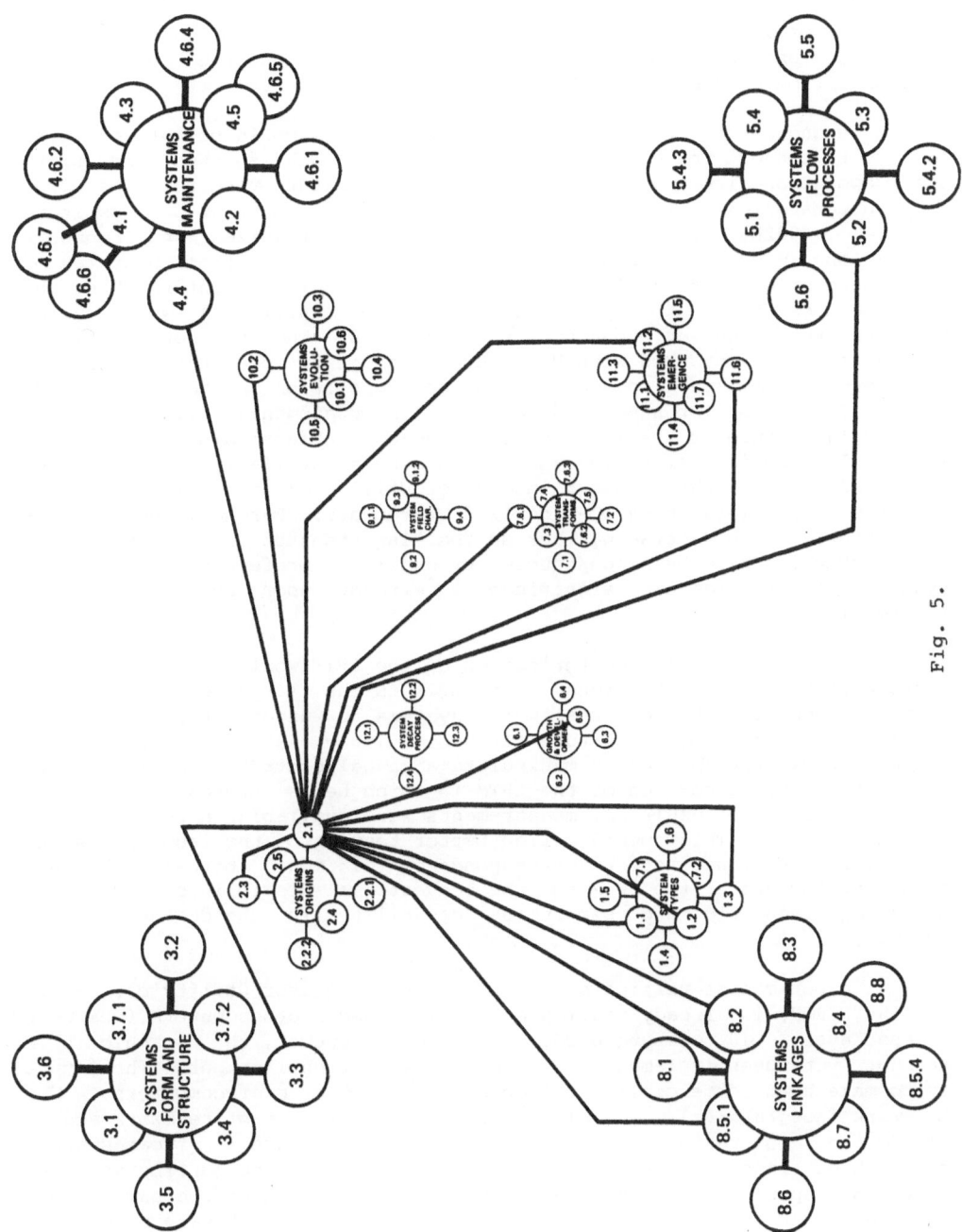

Fig. 5.

* Give me all Linkage Propositions using the following keywords in the statement.

* Show me a "time series" of LPs based on the generalized systems life-cycle.

A number of additional commands would be possible limited only by the formula for understanding of natural language built into the system, its interpretation of the graphics, and the user's ingenuity.

A second benefit of computerizing the graphics would be the resulting ability to connect the graphic representation of each isomorphy and Linkage Proposition directly to its significant literature results. The Lifework Integrator(c) suite of programs under development at our Institute (Troncale, 1982b, Miller and Troncale, 1982) organizes and connects stand-alone, discrete, condensed statements of fact from the refereed literature on each of the isomorphies listed in Table 2. Interlocking the LPTM Graphics Utility to the Lifework Integrator Utility(c) would allow the user to issue an expanded set of commands. The user could employ a "mouse" such as that used with Macintosh graphics to move the cursor to any visible isomorphy sphere or Linkage Proposition in the graphic LPTM (Figure 4) and ask for support data in English statements culled from the systems literature or the literature on the conventional disciplines. The full bibliographic reference would then also be available for instantaneous display since the Lifework Integrator Utility(c) has all references linked to its hierarchically-organized statements. Further, the Lifework Integrator(c) allows "tracing" across the "interconnections" in its unique data bank which have been constructed by qualified professionals in other studies. This feature significantly extends user control of the information.

A third benefit of computerization of the LPTM would be eventual correction of the various portions of the LPTM to the large compilation of empirical measurements on real world systems now under construction in a research project at our Institute (Troncale, 1981a, 1982c). Just as described above the user could manipulate a manual control to move the cursor to the visible portion of the LPTM in which he is interested and inquire what types of empirical measurements were available in the data base for that item. This would allow better testing of the isomorphies and linkage propositions, better "correspondence relations" between the abstract statements of the LPTM and the conventional disciplines, and better potential diagnostic and simulation capability in the decades of work ahead.

An experimental version of the Lifework Integrator Utility(c), written in FORTH, is already available on CP/M-based microcomputers (testcase is the Kaypro - 10 megabyte harddisc system). However, the graphic LPTM has not yet been implemented on a microcomputer environment. The intent is to make both commercially available on portable microcomputers. The Lifework Integrator(c) can also be used for synthesis of teamwork and as an aid to cooperation between distant collaborators. "Team-shared" research outlines could be generated with portions of the outline assigned to various specialists on the team. Many minds working on the same common computerized LPTM would enrich its referrals to the literature, enhance the efficiency of teamwork, and expand its networks of linkage. The Utility also provides for an orderly version of "what-has-been-accomplished-to-date" which serves as a stable foundation allowing precise addition and placement of incremental improvements until better completeness is achieved. Here, then, would be a practical tool for transdisciplinary research.

5.3 Mathematical Formalization of the LPTM

The sheer number and complexity of interactions in the LPTM may be rendered more user friendly with graphics, and still more useful as a practical tool if data bases are linked to the graphics, but neither extension simplifies the LPTM as much or gives it as much generative power as the mathematical formalization of the LPs (Troncale and Voorhees, 1983). The semantic statements of the LPs which describe singular, stand-alone interactions, can be effectively generalized by invention of a symbol to represent the interaction. The entire history of physics and mathematics revolves around the increasingly effective condensation of what first where ideas of relation, to better and better statements of relation, to symbolization of the relation. Gradually, the precise functions and limits of the meaning of each symbol were more exactly defined with each iteration. The result was the compact formulae of science, and the elaborate mathematical rules for manipulation of the formulae to obtain new knowledge about formerly unknown relations. These are now so common, they are assumed. But recent history indicates that many new and potentially powerful symbol systems await discovery. This project attempts to transform the linkage propositions into a new symbol set in order to obtain the economy of expression and generative power of mathematics for this sub-field of systems science (Troncale and Voorhees, 1983).

There are a number of precedents in the systems literature which bear some similarity to linkage propositions. Discrimination betwen LPs and the "correspondence principles" described for the natural sciences by the philosophers of science (Schaffner, 1969), the linked set of systems definitions of Ackoff (1971), the "entailment networks" of Pask (1974), and the "cross-level hypotheses" of Miller (1978), are described in former papers (Troncale, 1978, 1982b, 1983). Only the first of these have been formalized, although Pask's work has been recently computerized.

Since the LPs are linguistic and logic-based statements of hypothetical or observed mutual influences, they fall in part under the general title of symbolic studies of relations. Warfield (1980) describes in condensed form the history of symbolization and formalization of logical relations from the earliest attempts by Aristotle, Leibnitz, Euler, and Venn to their development in our century by Pierce, Wiener, Kuratowski, and von Neumann. He notes that the majority of improvements in formalization of logical relations has occurred during the last two decades. To this has been added the series of papers on Interpretative Structural Modeling (Warfield, 1976b, 1977, 1979). The network graphs of ISM produced from semantic statements input by participants may bear some relation to the LPTM graphs such that exploration of the theorems and methods of extension of ISM symbolics could be fruitful in the attempt at formalization of the LPTM.

The process leading to initial formalization of the LPTM is as follows: (i) source fields are continually surveyed for putative isomorphies to add to the set in Table 3. (ii) application of a "modest" general morphological technique (Zwicky, 1969, Zwicky and Wilson, 1967) and further search of the literature is used to formulate new linkage propositions to add to the computerized data base, (iii) analysis of the LPs enables clustering them into equivalences classes of association based on the similarities and dissimilarities of semantic statements connecting the isomorphies, (iv) a single, abstract symbol is devised to represent each distinct equivalence class of association such that the symbol is defined as the "operator" or "relator" possessing the "function" of the semantic statement typical of that association class, (v) the association classes themselves are explored for higher levels of interactions or influence among themselves ("meta-functions") or for significant "limits" of their function. The result is a set of more compact malleable statements of the general form

$$"A" \quad * \quad "B"$$

isomorph A stands in relation "*" to isomorph B

...where in all cases (the transdisciplinary, non-scalar, invariant) relation "*" represents one of a definable set of linkage propositions such that

    *(a)   possesses properties a(1), a(2),..........a(n)
    *(b)   possesses properties b(1), b(2),..........b(n), etc.

and each set represents an association class of similar LPs. These compact statements would contain a distillation of the most important of the interrelationships presented in the systems literature and would also, inherently, possess rules for their own recombination. If the definition of some of the association classes already formulated proves true (Troncale, 1978, 1982b), then some of the operators of this formal system would be quite unique. Also new rules of recombination are implicit in the LP sets reflecting the peculiar holistic nature of the systems they model. Rather than having to endlessly explain why systems results and tools do not conform to the expected standards of the conventional sciences, this development might allow systems scientists to specify (in a formal way) new levels of logical relation which could never have been found in the conventional disciplines.

    The following performance criteria for the formalized LPTM are discussed in (Troncale and Voorhees, 1983). The LPTM must: (i) achieve simplicity of representation that unifies the interactions of a multitude of systems ranging from masses of 10 exp (-32) gms to 2 x 10 exp (49) gms yet describes discrete interactions that give rise to observable systems-level results, (ii) achieve easy translation to conventional disciplines by connection to observables, (iii) achieve graduated transition to mathematical formalism to avoid early fossilization, and to make use of such internal checks as self-consistency, (iv) prove useful for generating additional linkage propositions in the manner of the following oversimplified example,

    (known LP)............A *(a) B
    Feedback(A) is necessary for Cycling(B) to occur.

    (known LP)............B *(k) C
    Cycling(B) partially contributes to Temporal Systems Boundaries.

    (unknown LP)..........A ? C
    Therefore, Feedback is necessary for maintenance of Boundary Condition.

(v) provide aid in intuition- or metaphor-building as well as more formal symbolization, comparison, recombination, and interrelation of such knowledge, and (iv) achieve robustness, that is, the LPTM must yield more information than that put into it. It must have demonstrable primary productivity and generative power to avoid the trap of "empty formalism" which is rather common in the systems approach.

    The first paper of this series also briefly describes seven expected contributions of the formal LPTM which derive from its special techniques of representation. These include: (i) improved expression and actualization of the age-old concept of "context", "nesting", or systems environ-

ment, (ii) better expression and use of the systems concept of "entitation" introduced by neurophysiologist and co-founder of the SGSR, Gerard
(see also section 4.1 , no.6), (iii) an ability to precisely demonstrate
"category switching", or "facetism", (iv) an ability to express "threshold
summation" in systems terms, (v) an ability to demonstrate the meaning of
systems "resonance" and its formal relation to scale invariance, (vi) an
ability to discover new interrelatiorships through the technique of "duality extension", and finally the extrapolation of unknown observables from
known categories of observables.

Let me just introduce the last potential technique as an example. If
an isomorphy "A" has the set of known observables

$$[A(i) : i = 1, 2,.......,n]$$

but another isomorphy "B" has a set of observables only partially known

$$[B(k) : k = 1, 7, 9]$$

but it is known that A is related to B by a linkage proposition or linkage
propositions set such that A (i to n) *(r) B(k), then one could search all
"i", "r", and known "k" and the associated systems data bases for unknown
cases of "k" conformable to "i" by relation "r".

Even though such examples are rather primitive, it is perhaps clear
that the "formal version" of the LPTM possesses interesting potential and
promises interesting contributions.

5.4 Graph Theory and the LPTM

Figure 4 indicates that the LPTM has the properties of a formal
graph. It is a collection of line segments that connect a series of
points. The aspects of line length, line type, line curvature, line position, and point content are insignificant features of the model. The
"connectedness" is the important feature of the model. Graph theory
shares these initial properties. The large number of theorems already
devised in the formal study of graphs could be profitably mined for any
relations to use to the LPTM.

Another indicator of the possible utility of graph theory to the LPTM
is the wide range of applications shared by both. Graph theory has been
applied to systems as disparate as man-made electronic systems, models of
molecular and organismic physiology, ecosystems, game theory, communications networks, transportation systems, models of crystals, family trees,
sociograms in sociology, and organizational structure. The LPTM, as a
general systems model is also very broad in its applications since it pur-
ports to represent the invariant, non-scalar, transdisciplinary mechanics
of systems function. Their subject domains co-map; possibly their results will also. A considerable amount of effort has already gone into
the theorems of graph theory. LPTM stands to benefit. Perhaps the benefits will accrue in both directions.

In addition, we have already alluded to three features of the LPTM
that are also features of graph theory. Graph theory has a series of theorems defining and describing the "tree" structures inherent in graphs.
LPTM also has internal tree structures, not the least being the taxonomic
tree of isomorphies shown in Table 2 and discussed in Section 2.3. Each
element of this meta-taxonomy is the base of its own set of relations and
trees. It will be interesting to see if any insight emerges form comparing across these two formalizations vis a vis tree and network transformations. Graph theory also has a relationship with simplexes of mathemati-

cal topology. The LPTM is proposed as a manifestation of the thermo-dynamically most likely multipart interaction state due to minimal needs for energy, space, and time. In the most fundamental sciences of sub-atomic particle physics and exploration of cosmology these minimal states have led modern scientists directly into the mathematics of the field of topology. The forms therein apparently model the minimal states (otherwise such basic form would not exist in the universe). Thus, the LPTM as a foundational minimal state may be expected to involve topology at a very primitive level. Finally, the relationship of the LPTM to the history of symbolic logic beginning with the Greeks was mentioned above. Graph theory also has as its central methodology the manipulation of logi-cal statements.

Some important differences exist between graph theory and the LPTM. It has been maintained that despite the efforts expended in the cause of formalizing graph theory, no property has been discovered that is charac-teristic of all graphs except for their general definition of being com-posed of lines and interstices. This may be the result of the tremendous diversity of graphs from extremely simple cases of one-line segment graphs to immensely complicated graphs. The LPTM does not co-map with this entire domain since it is a very complex graph to begin with. So it may be expected to avoid this dearth of general properties. Related to this, it is interesting to note that the LPTM only models "mature" natural systems as pointed out above, and is not expected to be as useful for very simple man-made systems which do not have to develop, evolve, transform, and emerge. LPTM needs help on the problem. As if in answer to this need the central problem of graph theory has been described as the identi-fication of significant classes of subgraphs and formalizing the relations between these subgraphs. Since the LPTM does not, as yet, have the same formal abilities to identify subgraphs of the total, graph theory could be utilized then to help formalize the limits of a general systems model by defining "domain of applicability" more precisely. Not all differences between the LPTM and graph theory lead to mutual aid. For instance, there is a tendency in graph theory to stress the identification of subgraphs that are "linearly independent" and yet cover the entire graph. This emphasis runs counter to the four emphases in the LPTM: namely "mutual" causation; "network", not linear cause and effect; threshold summation; and the importance of "context". Perhaps LPTM may make some contributions to a more general graph theory in addition to benefitting from contributions from the field.

5.5 The LPTM as an Expert System

The unfortunately-named "expert system" is a computer with software containing a sufficiently detailed version of a segment of human knowledge (and the rules for its manipulation) that it can answer a limited set of questions about that knowledge. Its key elements are a knowledge data base (usually from the literature), and an inference engine that emulates the human logic process. In essence it is an attempt to model the inter-connections man has discovered about a limited knowledge domain. The con-nections with the LPTM should be obvious.

But more importantly, the rapidly advancing set of tools for building "rule-based" systems makes it feasible to attempt still another technique for making the LPTM more user friendly and utilitarian. Because of its structure the LPTM is a likely candidate for becoming a rule-based system to study systems and to aid in systems studies. Here the terminology would be a little strained because the LPTM Rule-Based System would be used "by" experts not to supplant experts. The typical user would be an expert in a particular system (for example, transportation) that wants to model his area more completely, or diagnose and explore possible solutions to a problem in the area, but who has no formal knowledge of advanced systems science. The LPTM Rule-Based System would then be a meta-expert for use by experts. Presumably even a worker who had spent years modeling a specific system could benefit because the knowledge of a specific area of human experience is not equal to the combined generalized and organized knowledge of general systems. Transdisciplinary models, like the LPTM, encompass and go beyond the many separated areas of specialized knowledge shown in Table 3.

The LPTM Rule-Based System (called GENSYS) would consist of four major components. Its "inference engine" would borrow from established versions with the important addition of operators from the Linkage Propositions that define functions not featured in traditional logic, but which are necessary in systems logic. Its "knowledge base" would consist of facts from the fields of systems analysis, systems theory, and general systems theory as linked to data from the traditional sciences as described in Section 5.2. Its "rule-base" would consist of the linkage propositions themselves. An additional unique component would be the "Abstraction Mapping Utility" which would help the user who is an expert in his own subject domain identify "correspondences" between the isomorphies of the general model and the particular manifestations of those isomorphies in his system. It would do this with the set of characteristics and criteria now being assembled for the isomorphy taxonomy of Table 2, and for presentation in a general systems textbook (Troncale, 1981b).

These initial "correspondences" would be established by a series of menu's that quiz the user. The computer would use the results of these queries (in real time) to construct tables which equate the co-mappings. It would then be able to present, in the terminology of the particular discipline, the Linkage Propositions between components of interest to the user. There could be an "estimator of probability" as in some expert systems, but due to the hypothetical nature of many of the linkage propositions, probabilities in a chain of linkages would be very low. The use of the system would be more aimed at diagnosis of problems by suggesting impacts to the user, or as a design aid for identifying possible avenues of solution rather than as a provider of authoritative answers.

An extension of GENSYS would be META-GENSYS. This program would provide a service to the general systems science community. It would catalogue the work of the field, and provide a common data base for identification of work-in-progress, needed areas of work, identification of collaborators, and identification of ways to improve GENSYS in somewhat the same manner as self-improving expert systems already demonstrated (see section 4.1, Nos.12 and 13).

Both GENSYS and META-GENSYS would be implemented on CP/M microcomputers using FORTH as the programming language. The utilities would be built on relational data base substructures integrated with the Lifework Integrator(c) (already programmed in FORTH).

CONCLUSION

It must be obvious that much more work is anticipated by this paper than has been accomplished to date. Teams of cooperating colleagues are assembling on each of the above task. These teams cover a wide span of disciplines and technologies. The Society for General Systems Research is organizing Special Integration Groups and computer conferencing data bases which will indirectly aid in the advancement of some of these projects.

It should also be clear that although none of the above five exten-
sions of the LPTM is totally dependent on the others, progress in any of
them will significantly aid progress on the others. This type of synergy
may increase the probability of completion of the entire set of projects.

Finally, I return to the theme of the paper and this book. Can a
knowledge of natural systems help in the design of man-made systems and
aid our understanding of complex systems? Surely the increased resolution
of processes, their greater specification and interconnection, the attempt
at more complete coverage of the isomorphies characteristic of the LPTM
should increase both the ideas and the tools we have for design of our own
systems. If it is true that the general systems model is one that has
been optimized over many billions of years by nature for her systems, then
it could yield some important insights on how we need to build our systems. Even the immature systems of man, which I have said do not fall entirely within the domain of the LPTM, may benefit. Humankind may choose
to make systems immature only because it refuses to invest the effort/time
to make them more mature and because it is blissfully ignorant of the benefits that would result. The "special knowledge" of a future general theory of systems may reduce such ignorance, and future practical tools like
the LPTM may reduce the effort/time necessary for man to design improved
human systems based on natural systems' wisdom.

LIST OF TABLES

Table 1. Comparing the speciality areas in systems science using a dozen
          characteristics, each one showing a gradation from systems
          analysis at one extreme of the spectrum to general systems
          research at the other. It is important to recognize that the
          intent of this table is not to suggest a hard and fast
          distinction between these three "approaches" to systems science,
          but rather to indicate that each emphasizes or favors one or
          another of the extremes of the range of possibility for each of
          the 11 criteria.

SYSTEMS ANALYSIS	SYSTEMS THEORY	GENERAL THEORY OF SYSTEMS
Phenomenon-based	Discipline-based	Transdisciplinary (Inter, Multi)
Proximal to Data	Abstracted from Data	Abstracted from Theory
Predictive	Explanation	Understanding at Systems-Level
Not Isomorphy-based	Some Isomorphies	Connected Isomorphies (Many)
Most Specific	General	Most General
Correspondence Principles Apply ←————→		No Correspondence As Yet
Single Tool	More Than One Tool	Toolbox, The Source
Reductionist	Some Synthesis	AnaSynthetic
Mathematics and Computers ←————→		Conceptual to date
Field Insensitive	Field Insensitive	Field Sensitive
Less Contextual ←————————————→		Context Dependent

Table 2. Towards a comprehensive taxonomy of phenomenological
          isomorphies. A table of 75 principal systems concepts

1. TYPES AND TAXONOMIES
   1.1 Definition of Systems
   1.2 Parts / Components /
       Entities / Elements
   1.3 Purpose / Function /
       Equifinality / Determinism
       1.3.1 Externally-Generated
       1.3.2 Internally-Generated
   1.4 Subsystem / Supersystem
   1.5 Open Systems
   1.6 Closed Systems
   1.7 Types of Systems
       1.7.1 Decomposability
             (Fully, Nearly, Non)
       1.7.2 Linearity, etc.

2. SYSTEMS ORIGINS
   2.1 Boundary Conditions /
       Closure
   2.2 Self-Organization
       2.2.1 Autopoiesis
       2.2.2 Allopoiesis
   2.3 Self-Referential Mechanisms
   2.4 Non-Equilibrium
       Thermodynamics

3. SYSTEMS FORM / STRUCTURE
   3.1 Structureprocess
   3.2 Duality (Origins of)
   3.3 Hierarchical /
       Heterarchical Form
   3.4 Structure of Voids
   3.5 Fractal Structure
   3.6 Strings
   3.7 Correspondences
       3.7.1 Symmetry
       3.7.2 Asymmetry
       3.7.3 Supersymmetry

4. SYSTEMS MAINTENANCE
   4.1 Static States
   4.2 Stability
   4.3 Metastability
   4.4 Steady State / Dynamic
       Equilibrium
   4.5 Transtemporal Stability
   4.6 Control / Regulatory
       Mechanisms
       4.6.1 Negative Feedback
       4.6.2 Positive Feedback
       4.6.3 Coupled Feedback
       4.6.4 Feedforward
       4.6.5 2nd, 3rd Order
             Cybernetics
       4.6.6 Single Loop /
             Multiple Loop
             Feedback

       4.6.7 Hierarchical / Cross
             Level Feedback

5. SYSTEMS FLOW PROCESSES
   5.1 Flow Turbulence (Power
       Spectrum)
   5.2 Restructuring / Throughput
       / Temp. Capture
   5.3 Orthogenetic vs.
       Dispersive
   5.4 Energy-Based
       5.4.1 Entropic
       5.4.2 Negentropic
       5.4.3 Synergistic
   5.5 Information-Based
       5.5.1 Law of Requisite
             Variety
       5.5.2 Permutation /
             Recombination Mech.
   5.6 Optimality Principles
       5.6.1 Principle of Least
             Action / Energy
       5.6.2 Principle of Least
             Time / Space
       5.6.3 Principle of Least
             Matter / Energy

6. SYSTEMS GROWTH AND DEVELOPMENT
   6.1 von Baer's Laws
   6.2 Zipf's Law
   6.3 Morphometric Laws
   6.4 Allometric Growth
       (Proportionality)
   6.5 Principle of Plenitude

7. SYSTEMS TRANSFORMATION
   7.1 State Determined Systems
   7.2 Phases / States / Modes
   7.3 Catastrophe
   7.4 Bifurcations
   7.5 Cobordism Surgery
   7.6 Cyclical Behavior
       7.6.1 Life Cycles
       7.6.2 Limit Cycles
       7.6.3 Periodic /
             Oscillatory Behavior
   7.7 Thresholds

8. SYSTEMS LINKAGES
   8.1 System Context or
       Environment
   8.2 Input / Output
   8.3 Entitation
   8.4 Complexity Measures and
       Constraints

                              (continued)

Table 2 (continued).

8.5 Coupling Types
  8.5.1 Insulated / Non-Insulated
  8.5.2 Strong / Weak
  8.5.3 Synergistic / Antagonistic
  8.5.4 Linear / Non-Linear, etc.
  8.5.5 Internal / External
8.6 Coupling Magnitudes / Distances
8.7 Macro-Uncertainty Principle
8.8 Variety Measures: Diversity Measures

9. SYSTEMS FIELD CHARACTERISTICS
9.1 Resonance Phenomena
  9.1.1 Consonance
  9.1.2 Dissonance
  9.1.3 Transgressive Recursion
9.2 Soliton's (Long Waves)
9.3 Anticipatory / Precocious Vectors
9.4 Potential Spaces (Multidimensional)

10. SYSTEMS EVOLUTION
10.1 Randomness / Chaos Mechanisms
10.2 Concrescence
10.3 Neutrality Principle
10.4 Logarithmic Spiral of Variants
10.5 Transgressive Variation
10.6 Ontogenetic / Phylogenetic Mechanism
10.7 Lotka-Volterra Competition Equations
10.8 Cooperation Equations

11. SYSTEMS EMERGENCE
11.1 Stability Limits-Isomorph Networks
11.2 Parameter Trends
11.3 Process of Emergence
11.4 Satisfied vs. Unsatisfied / Counterparity
11.5 Transgressive Equilibrium
11.6 Exclusion Principle
11.7 Deutsch's Law

12. SYSTEMS DECAY PROCESSES
12.1 Instability
12.2 RECYCLING of Components
12.3 Programmed (Internally-Determined)
12.4 Externally-Determined

Table 3. All of these fields are useful for the detection and elucidation of systems isomorphies even if the intent of their research is not aimed at the discovery of systems concepts.

HOLISTIC INTELLECTUAL MOVEMENTS	APPLIED SYSTEMS ANALYSIS
Theoretical research	Applied research
Applicable to multiple disciplines	Applicable to multiple problems
Predominantly generative	Predominantly utilitarian
Some synthetic tendencies	Reductionistic emphasis
General Systems Theory	Management Systems Methods
Hierarchy Theory	Systems Theory in Ecology
Cybernetics	Political Systems Analysis
Information Theory	Defense Systems Analysis
Artificial Intelligence	Health Care Systems Analysis
General Topology	Energy Systems Analysis
Mathematical Systems Dynamics	Queuing Theory
Simulation and Modeling Theory	Game Theory
Synergetics	Operations Research
Global Modeling	Decision Theory
Systems Dynamics/Systems Engineering	Optimization Theory
Abstract Theoretical Physics and Cosmology	Graph Theory
Computer Systems Analysis	Technological Assessment
Information Theory	Systems Analysis in Behavioral and Social Sciences
General Morphology	Biological Systems Analysis

Table 4. Examples of Linkage Proposition

o Life cycles are a type of boundary condition, specifically defining temporal boundaries.
o Boundary conditions in part result from the establishment of a steady state, whether it be the result of either static or dynamic equilibrium
o Boundary conditions contribute in part to the cause of the exclusion principle.
o Boundary conditions of a system are in part the result of the strength aid duration of the linkages between its subsystems.
o The participations of entities / components / elements as subsystems in a supersystem is in part the cause of their transtemporal stability.
o Concrescence ratio can lead in part to the establishment of new boundaries.
o Transitions / phases / modes are in part the result of alterations in the linkages among subsystems of a system.
o Positive and negative feedback mechanisms are often found linked together / as a partial cause of dynamic equilibrium.
o Feedback is one of the few types of linkages which operates across widely separated levels of the hierarchy.
o Positive feedback / lrt / is a partial cause of amplification of rates of growth and development or decline and decay.
o Coupled positive and negative feedback mechanisms are in part the cause of oscillations around the ideal median.
o Transgressive equilibrium is in part the cause of levels in hierarchy.
o Recycling of systems components / after systems lifecycle death contributes in part to equilibrium of higher levels of the hierarchy.
o Instability and its opposite stability are paired in nature as a partial cause of one of the most fundamental of counterparities (dualism).
o Cycling reduces the energy flow necessary to maintain a negentropic, deterministic succession of states in a system.
o Goal-seeking feedback is in part the cause of oscillatory cycling.
o Metastability is a partial inhibitor of recycling of components.
o Reductions in required energy flow for cycling are partially dependent on contributions of recycling of components to autopoiesis of systems in succeeding hierarchical levels.
o A small amount of unsatisfied counterparities in a population of entities with mostly satisfied counterparities will result in concrescence and emergence of hierarchical structure.
o Coupled positive and negative feedback mechanisms are a generic example in counterparity.
o Neutrality quest is in part the result of the universal trend toward entropy death.
o Hierarchical organization is highly negentropic.
o Hierarchical organization contributes to systemic growth and development and allowable complexity limits.
o Gaps in hierarchical levels are the result of the appearance of new magnitudes of bonding strength, distance, time, and energy due to the appearance of new unsatisfied counterparities.
o Synergy contributes to negative entropy.
o Synergy is a special relationhip of input / output progresses such that the components sharing the relationship have achieved an unusual focusing of their outputs on each other as stimulatory input.
o Synergy maximizes temporal capture of enrgy flux.
o Open systems can locally increase their order of negentropy if energy is constantly supplied for throughput.
o Energy required for maintenance is proportional to the negentropy of a system: (Odum and Odum, 1976).
o Energy flows derive from counterparities seeking their complement to achieve a neutrality balance.

REFERENCES

Ackoff, R.L., 1971, Towards a system of systems concepts, Mgmt.Sci., 17: 661-671.

Ashby, W.R., 1963, "An Introduction to Cybernetics", John Wiley and Sons, N.Y.

Beer, S., 1972, "Brain of the Firm: The Managerial Cybernetics of Organization", Allen Lane Penguin Press, London.

Berlinski, D.D., 1976, "On Systems Analysis: An Essay Concerning the Limitations of Some Mathematical Methods", MIT Press, Cambridge.

von Bertalanffy, L., 1968, "General Systems Theory: Revised Edition", G.Braziller, N.Y.

Boffey, P.M., 1967, Systems analysis: No panacea for nation's domestic problems, Science, 158: 1028-1030.

Boulding, K.E., 1978, "Ecodynamics: A New Theory of Societal Evolution", Sage Publications, Beverly Hills.

Bronowski, J., 1978, "Magic, Science, and Civilization", Bamptom Lectures in America, Nr.20, Columbia University Press.

Bunge, M., 1959, "Metascientific Queries", Charles C. Thomas Press, Springfield.

Bunge, M., 1972, Metatheory, in: "Scientific Thought: Some Underlying Concepts, Methods, and Procedure". UNESCO, The Hague, and Paris.

Capra, F., 1975, "The Tao of Physics: An Exploration of the Parallels Between Modern Physics and Eastern Mysticism", Shambala Press, Boulder, Colorado.

Checkland, P., 1981, "Systems Thinking, Systems Practice", John Wiley, Chichester, England.

Chung-Yuan, C. (trans.), 1969, "Original Teachings of Ch'an Buddhism", Grove Press, N.Y.

Churchman, C.W., 1968, "The Systems Approach", Decalorte Press, N.Y.

Deutsch, K., 1966, "The Nerves of Government: Models of Political Communication and Control", Free Press, N.Y.

Easton, D., 1965, "A Systems Analysis of Political Life", John Wiley, N.Y.

Eigen, M., and Schuster, P., 1979, "The Hypercycle: A Principle of Natural Self-Organization", Springer-Verlag, Berlin.

Enns, R.H., 1980, "Non-Linear Phenomena in Physics and Biology: NATO Advanced Study Institute", Series B. Vol.75, Plenum Press, N.Y.

von Foerster, H., 1974, "The Cybernetics of Cybernetics", Univ. of Illinois Press, Urbana.

Gesyps, R.G., Klir, G.J., and Rogers, G., 1977, "Basic and Applied General Systems Research: A Bibliography", State Univ. of N.Y. at Binghamton.

Hai, A., and Klir, G.J., 1984, "An Empirical Investigation of Reconstructability Analysis: Part I. Probabalistic Systems", N.S.F. Grant Report ECS-8217103.

Haken, H., ed., 1977, "Synergetics", B.G. Teubner, Stuttgart.

Haken, H., ed., 1980, "Synergetics", Springer-Verlag, Heidelberg and Berlin.

Haken, H., ed., 1983, "Advanced Synergetics", Springer-Verlag, Heidelberg and Berlin.

Hoos, I.R., 1983, "Systems Analysis in Public Policy: A Critique", Univ. of California Press, Berkeley.

Iberall, A.S., 1972, "Toward a General Science of Viable Systems", McGraw-Hill, N.Y.

Jain, V., 1981, Structural analysis of general systems theory, Behav. Sci., 26: 51-62.

Jantsch, E., ed., 1980, "The Self-Organizing Universe: Scientific and Human Implications of the Emerging Paradigm of Evolution", Pergamon Press, Oxford.

Jantsch, E., ed., 1981, "The Evolutionary Vision: Toward a Unifying Paradigm of Physical, Biological, and Sociocultural Evolution", Westview Press, Boulder.

Klir, G.J., 1969, "An Approach to General Systems Theory", Van Nostrand Reinhold, N.Y.

Klir, G.J., 1984, "Architecture of Systems Problem Solving", John Wiley, N.Y.

Koestler, A., and Smythies, J.R., 1969, "Beyond Reductionism", Hutchinson, London.

Majone, G., and Quade, E.S., eds., 1984, "Pitfalls of Analysis", John Wiley, Chichester, England.

Mandelbrot, B.B., 1977, "Fractals: Form, Chance, and Dimension", W.H. Freeman, San Francisco.

Maturana, H.R., and Varela, F., 1980, "Autopoiesis and Cognition: The Realization of the Living", D. Reidel, Boston.

Merton, T., 1965, "The Way of Chuang-tzu", New Directions Publ.Corp., N.Y.

Miller, H., and Troncale, L.R., 1982, "The Lifework Integrator Program", Copyright, Santa Barbara, Ca.

Miller, J.G., 1978, "Living Systems", McGraw-Hill, N.Y.

Lucretius, 1st cent. A.D., "De Rerum Novarum."

Oren, T.I., 1985, I. Simulation methodology: A top-down view. II. Simulation: A taxonomy. III. Simulation models: a taxonomy. in: "Encyclopedia of Systems and Control", E.M. Singh, ed., Pergamon Press, Oxford.

Parsons, T., 1971, "The System of Modern Societies", Prentice-Hall, Englewood Cliffs.

Pask, G., 1974, "Conversation, Cognition, and Learning", Elsevier, Amsterdam.

Popper, K.R., 1959, "The Logic of Scientific Discovery", Basic Books, N.Y.

Prigogine, I., 1980, "From Being to Becoming: Time and Complexity in the Physical Sciences", W.H. Freeman, San Francisco.

Rapoport, A., 1968, Some system approaches to political theory, in: Varieties of Political Theory, Prentice-Hall, Englewood Cliffs.

Robbins, S., and Oliva, T., 1982a, "An empirical classification of general systems Theory Concepts", in: "A General Survey of Systems Methodology", L.Troncale, ed., Intersystems Publications, Seaside, Ca., 3-14.

Robbins, S., and Oliva, T.A., The empirical identification of fifty-one core general systems theory vocabulary components, 1982b, Behav. Sci., 27: 377-386.

Robbins, S., and Oliva, T.A., 1984, Usage of GST core concepts by discipline type, time period, and publication category, Behav. Sci., 29: 28-39.

Rosen, R., 1970, "Dynamical Systems Theory in Biology. Vol. I. Stability Theory And Its Applications", John Wiley, N.Y.

Schaffner, K.F., 1969, Correspondence rules, Philos.of Sci., 36:280.

Simon, H.A., 1969, "The Sciences of the Artificial", MIT Press, Cambridge.

Thom, R., 1975, "Structural Stability and Morphogenesis", W.A. Benjamin, Reading, Mass.

Trappl, R., Horn, W., and Klir, G.J., 1985, "Basic and Applied General Systems Research: A Bibliography, 1977-1984", Hemisphere, Washington, D.C.

Troncale, L.R., 1976, "Four Models for Training Environmental Education Teachers Using the General Systems Paradigm: First, Second, Third, and Interim Report", H.E.W. Contract No. 300-75-0224, Institute for Advanced Systems Studies, Cal-Poly, Pomona, Ca.

Troncale, L.R., 1978, Linkage propositions between fifty principal systems concepts, in: "Applied General Systems Research: Recent Developments and Trends", G.J. Klir, ed., Plenum Press, N.Y.

Troncale, L.R., 1981a, Are levels of complexity in bio-systems real? Applications of clustering theory to modeling systems emergence, in: "Applied Systems and Cybernetics. Vol. II. Systems Concepts, Models, and Methodology", G.E. Lasker, ed., Pergamon Press, N.Y.

Troncale, L.R., 1981b, "Nature's Enduring Patterns: A Survey of System's Concepts", Institute for Advanced Systems Studies, Cal-Poly, Pomona, Ca., (new edition in progress).

Troncale. L.R., ed., 1982a, "A General Survey of Systems Methodology: Volume One: Conceptual and Mathematical Tools; Volume Two: Applications To Real Systems", Intersystems Publications, Seaside, Ca., (see the Preface and Introduction).

Troncale, L.R., 1982b, Linkage propositions between principal systems concepts, in: "A General Survey of Systems Methodology", L. Troncale, ed., Intersystems, Seaside, Ca.

Troncale, L.R., 1982c, Testing hierarchy models with data using computerized, empirical data bases, in: "A General Survey of Systems Methodology", L.Troncale, ed., Intersystems, Seaside.

Troncale, L.R., 1984a, A hybrid systems method: Tests for hierarchy and links between isomorphs, in: "Progress in Cybernetics and Systems Research 2", R.Trappl, ed., North Holland, Amsterdam.

Troncale, L.R., 1984b, "The tao of systems science: systems science of the tao", Institute for Advanced Systems Studies, Cal-Poly, Pomona, Ca.

Troncale, L.R., 1985a, Future of General Systems Science: Obstacles, Potentials, Case Studies, Systems Research J., 2:43-84.

Troncale, L.R., 1985b, Duality Theory As An Isomorphy, in: "Proceedings of the 29th Annual Meeting of the Society For General Systems Research", B.Banathy, ed.

Troncale, L.R., and Voorhees, B., 1983, Towards a formalization of systems linkage propositions, in: "The Relation Between Major World Problems and Systems Learning", G.Lasker, ed., Intersystems, Seaside.

Troncale, L.R., and Wilson, A.G., 1977, Process-Orientation of Natural Systems versus Goal-Orientation of Man-Made Systems - a symposium, SGSR-AAAS Annual Meeting, Denver, Colorado. See abstract in "The General Systems Paradigm: Science of Change and Change of Science", J.White, ed., Proceed. 21st Annual Mtg., Publ. by Society for General Systems Research, p.164.

Union of International Associations, Integrative, unitary, and transdisciplinary concepts, 1976, in: "Yearbook of World Problems and Human Potential", U.I.A., Brussels, Belgium, Sect. K.

Waddington, C.H., 1977, "Tools for Thought", Basic Books, N.Y.

Warfield, J.N., 1976a, "Societal Systems: Planning, Policy, and Complexity", John Wiley, N.Y.

Warfield, J.N., 1976b, Implication Structures for System Interconnection Matrices, "IEEE Trans. on Systems, Man, and Cybernetics", SMC-6(1), 18-24.

Warfield, J.N., 1977, Crossing Theory and Hierarchy Mapping, "IEEE Trans. on Systems, Man, and Cybernetics", SMC-7(7), 505-523.

Warfield, J.N., 1979, Some Principles of Knowledge Organization, "IEEE Trans. on Systems, Man, and Cybernetics", SMC-9(6), 317-325.

Warfield, J.N., 1980, Complementary Relations and Map Reading, "IEEE Trans. on Systems, Man, and Cybernetics", SMC-10(6), 285-291.

Whyte, L.L., Wilson, A., and Wilson, D., eds., 1969, "Hierarchical Struc-
    tures", American Elsevier, N.Y.

Wiener, N., 1948, "Cybernetics", John Wiley, N.Y.

Wilber, K., ed., 1982, "The Holographic Paradigm and Other Paradoxes",
    Shambala Press, Boulder, Colorado.

Young, O.R., 1964, A survey of general systems theory, Genl.Syst.Yearbook,
    61-82.

Zabusky, N., and Kruskal, M., 1965, Solitons, Phys.Rev.Lett., 15:240.

Zwicky, F., 1969, "Discovery, Invention, and Research", Macmillan, N.Y.

Zwicky, F., and Wilson, A., 1967, "New Methods of Thought and Procedure",
    Springer-Verlag, Berlin.

GUIDELINES FOR INFLUENCING SOCIAL POLICY THROUGH STRATEGIC COMPUTER

SIMULATION MODELS

Dennis L. Meadows

INTRODUCTION

My university studies first started in the field of chemistry. However, after working for half a year as a student intern in a chromato-
graphic laboratory, I concluded that I did not wish to pursue a career in
that field. I began then to search for alternatives, and I soon discovered at MIT a curriculum within the management school based on a new computer modeling technique, 'system dynamics.' The group developing this technique was led by Professor Jay W. Forrester, and I went to MIT to earn my
PhD under his direction. After completing my degree, I joined the faculty
at MIT. There my wife, Dr. Donella Meadows, and I directed a major
international project that used system dynamics techniques to create a
model that clarified the long-term implications of physical and de-
mographic growth on the planet.

We left MIT in 1972 and moved to our present university, Dartmouth
College, where we have been involved in computer simulation and in model-
based consulting in three ways:

First, Donella has created a new undergraduate curriculum in the use
of modeling and systems analysis techniques.

Second, I have developed a graduate program which offers a Master of
Science and a doctorate in the same general area, the MS in Resource Systems and Policy Design. Third, Donella and I together created and still
direct the Resource Policy Center (RPC), a research institute which grew
during its first decade to include 45 people and to conduct nearly one
million dollars of contract research each year. Our institute's specialty
is creating computer models that will be useful for somebody, typically
public officials responsible for energy and natural management.

Although the RPC institute staff members are very pleasant people,
individuals and organizations will not give us money for these models, if
the models are not useful for some purpose. So there has been a strong
and constant pressure on all of us in the Resource Policy Center over the
past twelve years to make useful models. Indeed, we have developed a set
of guidelines that we use in each project to ensure that the results will
be useful. Those guidelines will be the basis for my talk this evening.
My remarks will follow the outline shown in Figure 1.

First I will define implementation, which takes two forms, depending
on whether it involves tactical or strategic models. Then I will give you

a very brief summary about the current state of affairs regarding implementation. Mainly, I will focus on strategic models, models of long-term social issues, although many of my points apply also to tactical models. Then I will talk about the socio-economic context for strategic analysis. I will next mention two ways that these models actually can be implemented in practice. To conclude this presentation, I had thought of describing in some detail one of our successful modeling projects.

Within the RPC we built, at a cost of over one million dollars, a long-range, energy price forecasting model, which is still being used by the US Department of Energy. But instead of talking about a past study, I will put my ideas and guidelines for achieving implementation into the form of suggestions for a group in Austria that might wish to do a long-term strategic model related to acid rain and forest damage. This will make it easier for you to test the generality of my rules; I will present my guidelines, and you can judge whether my approach would be appropriate in Austria.

I believe the general approach would be useful. The technique that we used to ensure clients of our energy model has been used in many different projects, ranging from a few thousand dollars up to a million dollars and focusing on topics as diverse as national energy policy and personal alcohol addiction. It is a relatively general technique. I believe it is also suited to the culture of Austria. For purposes of this exposition, I will pretend that all of us in this room are starting tonight on a three year effort to develop a strategic computer model that will eventually have a major influence on those who are working to develop appropriate responses to the phenomena of acid rain and forest damage.

1. TACTICAL VS. STRATEGIC MODELS

First let me describe the differences between tactical and strategic models. Implementation has a different form for each of the two types.

A tactical model is one which bears on some specific decision, often a decision which is made in a repeated cycle under fairly narrowly defined conditions. Suppose, for example, that we are operating a bakery and need each day to make a variety of different kinds of bread. We would need to take into account the cost of flour, the projected demand for different types of bread, and so forth. That would be a situation which lends itself to tactical modeling. There are a variety of different kinds of mathematical methods for building such models - models related to sales forecasts, to inventory control, to least-cost ways of formulating the bread.

1. Definition of Implementation
   tactical vs. strategic models

2. Current State of Affairs

3. Context of Analysis

4. Mechanism for Model Implementation

5. Case Study of a Successful Modeling
   Project

Fig. 1. Outline of the Talk

Methods for constructing and using tactical models constitute a fairly well-developed area in the field of operations research. There are excellent professional societies, whose members work in this area. While there are still very many improvements to be made, the level of quality and the degree of implementation for these tactical models is certainly very much greater than for the average strategic model.

Unfortunately, we would not be able to use a tactical model in our proposed study for Austrian forest damage. Clearly, there is little about forest damage which lends itself now to tactical modeling. Perhaps if we had found a piece of forested land that has already very many dead trees on it, there might be some interesting tactical modeling to be done. We could use such models to identify the least-cost way to move the dead trees off the land. But of course that model would not influence society's response to the overall problem of dying forests.

To incorporate the factors interesting to society at large, we would have to develop a strategic model. Even though there are typically very great problems in implementing strategic models, such models have important roles.

Before starting on a stratgic modeling effort we would have to understand clearly what exactly those functions are. Many people - not the ones in this room - but many people somehow have the impression that modeling is only done in order to predict the future. For tactical modeling that is sometimes the case, for strategic modeling it is almost never possible. The system being modeled is also subject to outside influences; the relationships are too uncertain to make any kind of predictive exercise very useful. But there are really important functions to these models; they are listed in Figure 2.

Strategic models can be very helpful to us in conducting research on the system itself. They provide a conceptual framework, which can provide an efficient basis for integrating diverse data on the issue. In the case of forest damage an appropriate modeling effort could be an extraordinarily efficient way to improve our understanding about the problems. They can play many other roles in research as well. First of all, even as you begin creating a model, you are forced to develop an unambiguous and clear understanding of the problem.

If you listen to many different people talking about forest damage in Austria, you will see that each of them has a slightly different concern.

1. RESEARCH

- establish the problem
- identify key factors
- point out inconsistencies
- determine priorities
- quantify future behaviors
- establish range of uncertainty

2. COMMUNICATION

- one way
- interactive

3. LEGITIMATION

4. EDUCATION

Fig. 2. Roles of Computer Models

Some are worried that people will be put out of work in the recreation industries, or that paper companies will have to shut down. Some are worried about soil damage, others may be worried about the survival of particular animal species. And there are industrialists who are concerned about this problem, because they think that society may make some "stupid" - or not very well based - decisions about how much industry should pay to reduce the emissions that may cause acid rain.

Most social problems have many different symptoms, and these will often confuse individuals about what is cause and what is effect. Building a model forces us to agree on at least one aspect that we wish to address, that's always a key step in achieving understanding.

Even though we have a very poor understanding of the system, strategic models can help us identify the key factors and see which factors are not so important, so that we need not pay attention to them. Unfortunately, there is a pronounced tendency with cases like forest damage for many agencies to go out and start gathering a lot of data, assuming that when society has many rooms full of data, someone can and will go through the data and understand what factors are important. But this is a very inefficient way to do research.

For one thing, it is generally the case that those agencies spend 85 percent of their time getting data on those things that make only 15 percent of difference. They measure things that are easy to measure, not necessarily the ones that are important. However, systematic sensitivity analyses conducted on a good strategic model can help us to understand where the key uncertainties are and where we should focus our energies in gathering new data.

Models help to point out inconsistencies. When you have to reduce everything down to mathematical equations, it is more difficult for people to hold two different viewpoints about the same cause-effect mechanism, because they have to agree on what goes into the model.

We cannot use these models to predict the future. But with them we can begin to understand the shape of several different possible futures and to understand which outcomes are impossible. I believe politics is often the art of arguing about which of several impossible outcomes is most attractive. If a model can tell us just to forget certain impossible options and to concentrate on the others, then it has made politics more efficient.

Models do have another purpose; they can provide a useful tool for communication. For example, I can often express my thoughts about reality by pointing to similarities or differences between the real problem and some model both I and my audience understand.

Simulations of models can provide a rich set of dynamic outcomes that would be hard to describe verbally. Models, and this would be very important for our projected forest study, can also serve to legitimate a model builder, to establish in the eyes of others that he is a worthy participant in a certain debate. There are many different ways to become legitimate (that is to become important, a member of the inner circle). If you are rich, then whatever you have to say on any issue is often automatically assumed to be important. Or if you have good friends with high positions in the government, you can be stupid and poor, but your opinion may still be considered important. Senior industrialists tend to be important on certain matters. People like us may not have any of these possibilities to be legitimated, but we can carry out a very good computer modeling study which becomes a basis for important conclusions in the de-

bate and which, therefore, certifies that we are legitimate participants in the particular discussion addressed by the model. And finally, once we know something useful, computer models can be used for educational purposes. They can help us to tell others what we learned.

Implementation of strategic models is attained when you carry out a model project in a way that achieves one or more of these objectives. Notice, I did not say that the purpose is to publish a paper in an academic journal; nor did I say that the purpose is to generate 15 feet of computer printout or that the purpose is to make a certain number accurate to 5 decimal places. These goals are commonly observed in the modeling process, but there is hardly any relation to the strategic decision making process.

When you have pointed out inconsistencies, determined priorities, identified the key factors, and established the problem, then you can begin to define strategic models which actually come to have some influence on day-to-day decision making.

Unfortunately, most strategic modeling efforts never come that far. They have suffered from a set of problems listed in Figure 3. They are unimportant, because they deal with issues very interesting to the modelers, but not interesting to anybody else. Many people build models, saying, 'First I will study something that I am interested in. When I am finished, I will go find a client who can use the model.' This is a certain recipe for failure.

Today's models are usually unbalanced. They are influenced by one discipline, and thus they leave out factors from other disciplines that are critically important for understanding real policy issues. For example, if you had a group of ecologists make a 'forest damage' model, you would have the little microorganisms represented in it, but you would not have any of the big economic pressures that are also very important. On the other hand, economists' models of environmental issues typically ignore the last 50 years of progress in the life sciences. When you have an unbalanced model, it's like a car with only one wheel, it doesn't make any difference how superb, or how expensive that wheel is. If there is only one wheel on the car, the car is not going anywhere. Most models experience analogous results.

To be concrete on this point I can cite the experience of two scientists within my project at IIASA. Recently they conducted a world-wide survey, involving hundreds of scientists, to identify "ecological-economic" models. They sought models that had balanced representations of economic and ecological factors governing environmental issues. They did not find more than a few, much less than 10, which even came close.

Models have typically been:

1. Unimportant

2. Unbalanced

3. Undocumented

4. Unfinished

Fig. 3. Problems with Past Models

Models are very poorly documented. A common attitude of many model-ers seems to be, 'Well, I know everything about the model, just ask me, if you wish to discover how I produced any of my results.' In most cases you have no choice except to ask them, because it is impossible to guess from any written sources just how they derived their published results. It has been estimated, and my own experience confirms this, that adequate docu-mentation of a model requires just as much time as building the model. Most modelers never make that effort. This is an extremely important fact, when we consider factors governing implementation.

Documentation imposes a discipline on an analyst that can substan-tially raise the quality, even of a project that has been honestly and competently carried out. If you know that somebody else is going to be able to reproduce your results, or will try but fail to reproduce your results, it imposes on you more attention to detail than if you think that nobody will ever be able to simulate your models, unless you are present. The situation with most modelers today is analogous to being the teller in a bank where there is no accountant. Again data from my own experience can make this point concrete.

I conduct a graduate training seminar, in which each of my students goes through the literature and finds one mathematical model on some issue, normally in the area of agricultural economics. This is a useful field for testing documentation, because agricultural economics is the field in which social system modeling is most developed. There are very good data in agriculture, and the economic theories are the best-developed in the agricultural part of economics. Moreover, there are numerous well-established journals with relatively high scientific standards. Thus, if you just want to find a good model, and you do not care about the topic, you go to agricultural economics literature.

Over the years my students and I have taken about 20 of these models and examined them in great detail. We put the equations on our computer and run them. We looked up the data-sources cited in the paper. We ran the models and tested their sensitivity to modest changes in assumptions. We found that even in this field, the best area of mathematical modeling in the social sciences, at least 85 percent of the models are not docu-mented in a way which lets anybody else reproduce the published results! And of those 85 percent, in many cases the person who wrote the paper could no longer reproduce the published results, even working from personal records that are unavailable to others.

A detailed list of the reasons for these failures would make an in-teresting talk by itself. Let me merely summarize the factors here very briefly, because they are problems that you would have to avoid in your forest study, if anyone is to use your results.

Software mistakes are a problem. In one case that we studied, the author changed from one FORTRAN-compiler to another in the middle of the study. There was a mistake in the first compiler, so that it produced results which could not be reproduced with the second. No one noticed the mistake at the time, and now it is impossible to recreate the numbers gen-erated by the first compiler.

Another common reason is typing mistakes. It is incredibly easy to misplace a decimal point when you are entering data to generate a model run. Since detailed records of each rerun are seldom kept, it is easy to generate a model output using inputs that no one will ever be able to rec-reate. We found several examples of that problem.

I have concluded that deceit is sometimes also the reason for poor documentation. To be precise, I think some authors of the papers we studied published conclusions, which he or she actually knew was incorrect. They had some reason to want specific published results, and they went through the motions of generating the results scientifically; actually it was just a fraud. Of course they did not provide full details on the study!

There have been 22 major, published, global modeling projects since 1972. For how many of these do you suppose an independent reader could reproduce the published results using the data and equations that are publicly available? Only 3 out of 22 have been documented in a way that would permit an independent analyst to reproduce them. This is not science!!

The last problem with models I will mention tonight is that they are often unfinished. Why? In a typical modeling project the first 95 percent of the time are spent trying to get the model to run for the first time. And then only 5 percent of the time remain to revise it, debug it, do sensitivity analysis, write up the final report, make presentations, and find funding for the next project. It is no wonder that most models never get truly finished. If we want our models to avoid these mistakes, we have to understand the reason for the current situation.

One reason is that many people who pay for strategic models are perfectly happy with this situation. They do not actually want to contribute to science; they want to support a modeling effort that will lend credibility to some specific conclusion or recommendation. If nobody can question that conclusion by examining the bases for the model results, these clients are even happier. So they pay an analyst, he produces the predetermined result, this result is used for political purposes, and neither party cares about documenting the model. Indeed, their purposes are better achieved, if no one can examine too closely the foundations for their published results.

Another reason is that the discipline of strategic modeling for social system policy makers is quite new. It has not yet matured to the point where it has a well-established set of criteria governing the work of those professionals who practice it. There are few professional training programs in this area, and the professional societies and journals related to strategic modeling are a diverse and incohesive lot. Worse, the glamour and funds that were available in support of large strategic modeling efforts in the 1970s often attracted into the field people whose training and professional experience were quite remote from that required for good work. And of course most of these 'immigrants' were not at the top ranks of their original field. If they had been, they would have remained in it. I suppose (and I hope) that the field will have matured within another decade, so that the sloppy standards that currently characterize much work in the field will have been raised significantly. However, even when that has occurred there will still be a serious obstacle to implementation of strategic models. Strategic analysis takes place in the context of an issue lifecycle that makes it extremely difficult to find money for research as early as it is needed. This lifecycle is shown in Figure 4.

To create a useful model of forest damage, for example, may easily require 3-5 years of hard research. Unfortunately money is seldom available 3 to 5 years before an issue becomes politically important. However, when an issue heats up politically, the decision makers can seldom wait several years for the information they will use in formulating their basic decisions. When people are finally concerned about, let's say, forests damage or the energy crisis, then there is money for modeling. But 3 years before that they are worried about something else, and there is no money.

Our coal modeling effort illustrated the problem. I went to Washington, D.C., in 1972 to find money for a new project on coal. I wanted to build a model that would show how to increase the production of coal in the USA In 1972 coal had been a steadily declining fraction of US energy supplies for decades, and I found only one man in Washington who considered our proposal to be interesting. General Lincoln was the head of the Office for Emergency Preparedness. But he was only a few months from retirement, so nobody cared what he thought.

When he left, no one in the United States Government cared much about programs to increase coal production, because everybody knew that the coal output would be declining indefinitely. And that was twelve months before the oil embargo! Of course, if you had gone around in Austria three years ago to raise money for a big project on the death of the Austrian forests, you would also have found little interest in funding your work.

As illustrated in Figure 4, public resources for modeling increase much later than public concern. But a strategic model can only have a real influence between stage one and two in the development of a strategic issue, and during that period there are hardly any public resources available. After stage two the issues are so political that most policy makers have made a commitment, and there is not much open-mindedness or flexibility left. You can get a lot of money for a project during stage 3 or 4, but by this time everybody knows already what they want to do, political commitments have been made, and it is impossible to have much basic influence on the strategic discussion.

To cope with this dilemma, we have adopted the following procedure. Find somehow the money required for the first model on some strategic issue, and get most of the work done before the issue reaches stage one. When the issue has become important, secure a large amount of money to pay for what you have already done. Use that money to lay the foundation for the next strategic issue, in which, however, for the moment nobody is interested. It seems to me that in Austria I would start the first effort within the university system, where people's salaries are guaranteed irrespective of what they work on.

Suppose for the moment that despite all the problems mentioned above, we successfully create a valid, balanced model, which is well documented,

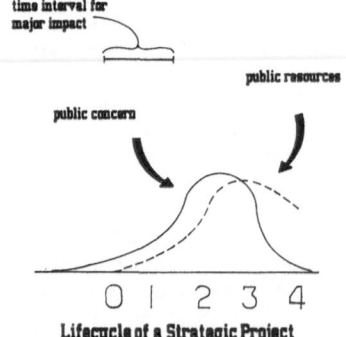

Fig. 4. Context of Analysis

related to an important issue, and completed in time to have some influence on the debate. What are the mechanisms for model implementation?

A conventional view of the implementation process has the analyst standing before a large group of decision makers with a massive pile of computer-generated charts. He spends fifteen minutes telling about the technical aspects of the computer model. Then, having established their validity, he shows a few computer runs, and indicates which policies produce the best outcomes. Everyone applauds; then they return to their offices and immediately start enacting the recommended policies. This is admittedly oversimplified, but it has most of the ingredients of the standard view about model implementation. And the process I just described has nothing to do with real implementation; it works only for the decision makers who were, for their own reasons, already committed to enacting the recommended policies. All others would just dismiss the analysis and recommendations, if they understood them in the first place.

Now, what can be the ways that strategic modeling ever does have some influence? There are two:

- your model may lead to a book, film, or report which shifts the balance of power slightly towards that faction that already supports the views you propound.

- your model may alter the perceptions, insights, opinions, and goals of someone who sits in the inner circle of decision makers that will influence the nature of the response of the issue of concern.

Great literary talent and substantial luck are generally required to produce a book that gains a widespread public office. They are available to a select, lucky few. We cannot count on them for the forest study. But the second means of implementation is one open to all of us. It is the one we should pursue in our proposed study of forest damage. It is the one I have used with consistent success in my institute.

I know of two ways for modeling projects to influence the opinions of those in the inner circle. One is to put the model in a form that makes it directly interesting and useful to people who are already members of the inner circle. The second is to conduct the study, so that one of the central members of the modeling team is helped to join the inner circle, as a result of the study.

During this past year I have been experimenting with a new technique for accomplishing the first outcome. I am now developing at the International Institute for Applied Systems Analysis some techniques that take sophisticated strategic computer models and build them into management training games. To be successful, this technique has to be carried out in a way that can interest senior managers in playing the games. After finishing the model, I build it into a multi-person game. In the process of learning how to win the game, managers also learn how to win in the real situation represented by the model. Managers never need to look at a mathematical equation, they don't have to evaluate my model directly, they evaluate the game and compare it with the real system. The preliminary results from this experiment are encouraging, but my conclusions from it are still tentative. Thus with my remaining time tonight, I will concentrate on procedures for accomplishing the second outcome; it is the approach we have used over the past decade within the Resource Policy Center.

I sit within a university. My colleagues and I in the RPC conduct each of our model studies deliberately in a way that makes one or more of our students into recognized experts. And each study is carried out systematically with this goal in mind. When the study is finished, the student who has done the work and has become an expert, will move to a job there - generally as an adviser to someone who sits in the inner circle. To produce that result requires that the entire project be carefully planned from the start to attain that goal. Therefore, implementation is not an activity to which one turns after all other phases of the project are completed! It is an orientation and a commitment that must profoundly influence every single stage of the analysis, from the initial conceptualization of the problem through to the delivery of the final report. Tonight I will present several guidelines for managing this process successfully, so that the model-based study eventually leads to concrete and constructive changes. My remarks are structured according to the eight phases of a successful modeling project; they are listed in Figure 5. I will discuss the eight areas and illustrate the application of our guidelines to a model on Austrian forest damage. Of course, I know very little about the mechanisms of forest damage. But the implementation process does not depend on the technical substance of the problem addressed by the model.

Identification of the issue

First comes the effort to provide a name and an orientation to the project. You have to name it. We know that the forests of Austria - and, of course, of other parts of Europe - now show very high damage levels. But we have to be careful. Our central question should have basically a positive slant, not a negative one. We should not, for example, characterize our study with the question, 'When will the forests in Austria be totally destroyed?' Few people could become very enthusiastic about such a study, even if they peronally have the opinion that forests are doomed. In our energy study we did not ask, 'When will oil depletion cause a crisis?' Our study sought instead to answer the query, 'What are ways to increase the production of coal with acceptable social, environmental and economic costs?' You have also to couch the issue in a way that means important groups have not already made up their mind about it. If you couched the issue: 'How much harm does the paper industry cause to the Austrian forests?', you would immediately have important people objecting to your research. They will try to block your funding. If you get money,

- identification of the issue

- institutional setting

- funding

- legitimation of the team
advisory council
publication policy
major survey conference
press relations

- recruiting partners
concept of the invisible college

- schedule for the work

- dissemination of results

- caveat about standards

Fig. 5. Case Study of a Successful Modeling Project

they will deny you the data you need, and so forth. Most important, come into the issue before there is much public concern - before the political attitudes have started to be hardened.

Perhaps your study could focus on tourism. Everybody understands that tourism is very important to Austria, and most generally favor it. You might choose as your issue the relationship between forest damage and tourism. "What long-term forest management policies will enhance the contribution of tourism to Austria's development?"

Institutional setting

Where is this project going to sit? If we want to conduct the study without secrecy in a location that permits many people to be involved, perhaps it could be conducted within the university. Alternatively, you might locate the study group in a consulting company. Perhaps some government ministry could take responsibilty. This would give you better access to certain types of important data, but this choice requires some care. It will be important to pursue the study consistently for three or four years. At least in the USA, there is so much turmoil in most political settings that could be difficult. Also, government bureaucracies have carefully partitioned the world and assigned each aspect of it to a specific department. Strategic issues almost always require one to deal with a set of phenomena that lie within the jurisdiction of many different government agencies. Thus the attempt by any one department to carry out a strategic modeling effort will often arouse opposition from other departments that has nothing to do with the quality or the relevance of the study.

Established institutions may already have prestige. They may have some excellent people; they have contacts; they have money. But on the other hand, they have some pressures on them. They have a lot to lose. They may have a past history of public pronouncements on the issue. If this is a study which is risky, you do not want the other interests of your host institution to influence how you are going to carry out the study. Perhaps you may decide to create a new institute; the Institute on Promotion of Sustainable Tourism.....or whatever. In any event, the choice of institutional setting deserves some really careful thought.

Funding

Generally the situation in the USA is much easier when it comes to funding. There are so many more potential funding sources in the USA In Europe, a single viewpoint, political orientation, or vested interest may characterize almost all the important funding sources for work on some specific issue. If your work seems to challenge this view, it may be difficult to find money for it. Fund raising is the phase of the project most influenced by the particulars of the institutional setting. Here I probably have little to offer my Austrian colleagues besides caveats.

One caveat: Modeling is expensive, so be certain to be realistic in assessing your needs. Remember that documentation takes as much effort and time as all the rest of the project together. One advantage you have here, so far as I can tell, is that university students and faculty tend to be supported irrespective of their work. Thus it may be best to get the forest study started within some appropriate academic group, which has its salaries guaranteed for a period long enough to complete the initial phases of the research.

Legitimation of the team

Then you have to start thinking about how to make your team legiti- mate, that is to say credible. Accuracy and expertise are required, of course, but they are typically not enough on strategic issues. In any e- vent you typically will not have much expertise on the technical details of the issue when you start your study. Therefore other means must be found to convince others that your work is worthwhile.

One way to establish the respectability of your work is to associate with it an advisory council of eminent people. There are different ways to get people to be on your advisory council, even if you are not famous. The best is to find one well-known and widely respected person who will join your group because you have convinced him that the problem is impor- tant and your group does have the capacity to help solve it. Then many other good people will typically join your council because of the first member's implied approval. Of course the council is not just for show. Use its members well and treat them with total honesty. Otherwise they naturally will not remain associated with your project.

A proper publications policy is also required, when you are working to enhance your credibility. Your reputation requires that you produce a stream of reports that are well-designed and rapidly available. Academic journals have very long publication delays, in the United States it may be nine months or more. Thus, at least in the early stages of the work do not use academic journals. At the RPC we distributed our own reports. We computerized our addresses and distribution system. That way we could exactly control the distribution. In that way we always knew who received our reports and who asked for copies. An extremely important issue in the early phase of the project is to know who has been expressing any interest in your work. These people may be the source of data or funding that is useful to you later. If you just give your reports to some other organi- zation, they will seldom be able to supply you with that information.

Remember that the physical appearance of your reports is used by many people to reach initial judgement about the quality of your work. We have a saying: 'You can't judge a book from its cover.' That is, you cannot tell if a book is worth reading from its cover. But the people of inter- est to you, those who constitute the inner circle related to your strate- gic issue, are overloaded with information from many sources. They will often decide not to read a report simply by looking at its cover. At the RPC we created our own report series. We hired a professional graphics designer to work out format, logo, cover design, and so forth. We sought to create a distinctive image through our report series. During the 3 - 5 year life of your project, maybe fifteen or twenty to fifty reports will come out. They should have a consistent form, their design should cause people to tend to put them on the bookshelf for storage rather than in the waste basket. After spending several tens of thousands of dollars to produce the contents of a report, you can afford to spend more that .50 on the cover! What holds true for the physical appearance of your report is doubly true for its literary style. Even a small team of five to ten people may find it worthwhile to have a full-time editor, a really excel- lent technical writer working with them.

A fourth avenue to legitimacy is through hosting an important inter- national conference on the topic. Identify the ten or fifteen people in the world who have something important to say on this issue. Bring them together for a survey conference. One of the most efficient ways to be- come fully informed about the current state of knowledge on an issue is to mix with those who are at the forefront of the field. If you show that you respect their work, ask good questions, and listen to the answers carefully, after a couple of days you will have something interesting to say yourself. The key thing is to get five people who are important and get them committed to come. Then most of the others will want to come also just to be able to visit with those five.

And start thinking about press relations. Your goal is not so much to get attention to yourself, but to find a channel, so that when the important issues arise, and you have information that should get out, the press will pick it up quickly. Find two or three key publications that will have an important influence on the issue over a time, because they are read regularly by members of the inner circle. <u>Science</u>, <u>Technology Review</u>, and <u>The Washington Post</u> are examples of such periodicals in the United States. Find in each of these publications one key journalist who is willing to invest a little time now, learning from your work, so that three, six or eight months from now he will be better able to inform the public about the issue. Involve those people in the early stages of your project, and treat them as equal partners in the project. Invite them to your key meetings and send them prepublication copies of your reports. At some critical juncture in your study, a brief mention of your team and its results in an important daily or weekly news forum will be more important to you than three major reports. Make sure you are on good terms with the people who can give you that mention. During our coal project there came a point when we needed to disseminate widely a report we had made on the long-term effects of President Carter's energy proposals. We were able to have it published as the cover article in <u>Technology Review</u> almost immediately, because of the relationship of trust and respect that we had built up with the editors of that journal.

Recruiting partners

Academics, at least in the United States, often act as if success comes from a zero-sum game. If somebody else has a little more success, they seem to believe that less success will be left over for them. In my experience with strategic issues it is generally exactly the opposite. Either the members of a group can help each other all succeed, or they can all fail. Find 10 - 15 people who, you think, are on the right side of the issue; get into a partnership with them. Work in a way that promotes your joint success.

You need some people with money, some people who are good scientists, some people who can hire the analyst who will graduate out of your project. You need several people in the media and a few who are influential members of the business and financial community. So sit down and identify the different kinds of people that you want to find in Austria who would be appropriate partners for you. In recruiting them, the concept of the invisible college is very important.

From many social studies - we know that in any field of intellectual activity, there are ten or fifteen people who really are the information stars. By information star we mean the person in the center of a network, from which there are many communication channels going out. For your work on forest damage the following approach would work. Go to an important official in the forest ministry. Ask him a set of questions: "If you had a technical or political question about factors governing your efforts to assess and reduce forest damage, who would you first call for advice?, Who has written the best material in this area?, 'Who is making the important decisions?, Who is most important in shaping public opinion on this issue?"

Then go to the people mentioned in response to your questions and ask them the same questions. Compile all the names that are mentioned. Put all these names on a big piece of paper and start drawing lines to show who communicates with them. You will find that some people are connected to only one other person. Others are in the middle - they are connected to everybody, directly or indirectly. Ten to fifteen people will have the majority of the linkages; these are the members of the invisible college, the information stars. They are the people of principal interest to you as you identify your partners.

Once I wanted to do a piece of analysis that would change the law influencing alcohol addiction in the state of Maine. I did not know anything about Maine, and I did not know very much about alcohol addiction at that time. But I knew, that if my work was to have any lasting influence, I would need to know and to work with the members of the invisible college that governs alcohol policy in Maine. The process by which one identifies the invisible college is so standard, that I told my secretary how to do it. She also did not know anything about Maine, and she, for sure, did not know anything about alcohol addiction. I went away on a trip and I came back three weeks later. She had identified most of the members of this college, simply by making about 20 telephone calls.

When we started to do our coal study, I hired a college student and sent him to Washington for the summer. He was a business student. He did not know anything about coal, but dressed very well, and he was smart. I gave him money to take people out to lunch. And I told him the names of three people who were in the invisible college related to national coal policy. He took them out to lunch, and he asked them the standard questions, including 'Who are the people that your secretary puts through instead of telling them to call back?". And when he came back at the end of the summer, he had a list of about fifty people. We picked fifteen people from that list and most of those people became the advisory board for our study about coal.

Schedule for the work

As you have been carrying out the activities mentioned above, you have, of course, been starting the technical research. A fundamental requirement for success is the ability to develop a realistic schedule, and stick to it. For this there are a number of guidelines. One rule of thumb, worth mentioning a third time, is: Documentation takes as much time as everything else put together! Make sure that it is in your schedule. Another guideline is to identify some key event that will create a special interest in and receptivity to your results. Is there something coming six months, twelve months from now - an election, perhaps, or a law that will expire and force new parliamentary debate? Is there the likelihood of some accident or political calamity that will create a big, albeit brief, surge in public concern. When you identify such events, schedule your work so that when that event comes, you will have something which is useful to those concerned about it.

Dissemination of results

Most analysts believe that procedures for disseminating the results of their work can be ignored until after the research is completed. But a study of my remarks will show that almost everything I have said so far relates to dissemination of results. We think carefully about the process of distributing our conclusions to the inner circle even before we have decided precisely what questions we want to address.

Caveat about standards

Now, what about the science in all this? Everything that I have said so far has no meaning, no integrity, if you do not conduct your study at the highest standards of honesty and excellence. If you actually do not have any expertise about the problem, or you are a lousy modeler, or you do not use good statistics, or you cannot write, then your project is not going to have any useful effects. And all of the things I have discussed so far will only call a lot of attention to that fact. All the rest of this business is like a carnival, if there is not a solid core of integrity and high quality research beneath it all, and if you are not always being honest.

But the opposite is also true. You can be a superb scientist; you can have wonderful knowledge about the problem; you can develop some extremely creative and effective new solutions to strategic problems. But if you do not carry out your work from the first day with a concern for its ultimate implementation:

NO ONE WILL EVER KNOW!

Of course, the guidelines that I have discussed do not guarantee success. At my institute we might start five or ten strategic projects for every model which really goes all the way to completion and to widespread implementation. But if one has a clear understanding of the guidelines and constantly monitors each project to determine whether the guidelines are being met, it is possible to cut short the projects that are not going to be implemented. Ask yourself constantly: "Do I have effective partners from the invisible college? Where am I on the public interest curve? Is there solid interest in our report series? Are we meeting our schedule targets? Is the scientific content of the work adequate?" When your answers to many of these questions begin to be "No!", stop the work, and start something else or redefine the project in a way that is more effective. If you can be really honest about your progress you can normally cut short your losses. At the RPC only one out of ten projects may succeed, but we end up spending at least eighty percent of our money on succesful projects.

Have confidence

I have saved the most important guideline for last. Always work from the context that your research can make a real difference to society. It is hard to believe, I know. But it really is possible even for a small group of people to have quite significant influence on major social issues. Remember, the public and corporate leaders who are confronted with strategic social issues desperately need the insights that can come from well executed modeling projects. But most people who are working in this field, build models that are unimportant, unbalanced, undocumented, and unfinished. As soon as your work avoids those four problems you are already one of the best analysts around.

REDUCING INTERNATIONAL TENSION AND IMPROVING MUTUAL UNDERSTANDING

THROUGH ARTIFICIAL INTELLIGENCE: 3 POTENTIAL APPROACHES

Robert Trappl

SUMMARY

This paper

- gives a short definition of Artificial Intelligence,
- explains briefly its main applications for commercial and especially military purposes, and
- shows its potential as a means to reduce international tension and to increase mutual understanding, especially in the situation of crises, by sketching 3 specific approaches.

WHAT IS ARTIFICIAL INTELLIGENCE?

Artificial Intelligence (AI) is the science and technique of making computers smart. Important areas of AI are knowledge representation, learning, reasoning and planning, language understanding and production, vision and sensing.

All important universities in the USA and many universities in Europe have established AI departments and offer AI curricula, leading in many cases already to a doctorate in AI. A large number of AI journals, text-books, and books in special AI disciplines are published every year. There are also several national and multinational scientific AI organizations.

A short introduction into AI is given in the "one-hour course" by Trappl (1986), more extensive are the books by Winston (1984) and by Charniak and McDermott (1985). The Handbook of Artificial Intelligence in three volumes gives the most in-depth presentation (Barr and Feigenbaum, 1981; Barr and Feigenbaum, 1982; Cohen and Feigenbaum, 1982).

AI FOR COMMERCIAL AND MILITARY PURPOSES

Already in the late seventies it became clear that AI results could be applied to practical problems:

- As intelligent assistants ("expert systems") when e.g. searching for the diagnosis of an uncommon disease, or interpreting the down-hole data from drilling sites, or recommending the purchase of stocks,
- for conversing with people, giving information from a data base or

translating from one language to another ("natural language systems"), and
- for working in factories, doing jobs in remote or dangerous areas ("robots").

With the advent of commercially successful products all major computer companies established AI departments; about 200 independent companies which develop only AI software have been founded. More than U.S.$ 100 million venture capital have been raised for them. In 1985, an estimated 150 companies, including General Electric, Gould, Shell Oil, and 3M, spent U.S.$ 1 billion to maintain in-house AI groups (Smith, 1985).

At the same time, several nations and multinational organizations started funding AI research, with sums going into billions of U.S.$. Among others, the EEC started the European Strategic Programme for Research and Development in Information Technologies (ESPRIT) with approximately U.S.$ 1.5 billion, the largest subproject of which is AI. While most of the projects in these programs are oriented to produce marketable products, it is quite clear that a considerable amount of the projects will strongly promote military developments.

In fact, the defense departments of most larger nations have already been funding AI research heavily for several years. The Defense Advanced Research Projects Agency (DARPA) of the U.S. Department of Defense has initiated the so-called Strategic Computing Program which aims at the development of autonomous land vehicles equipped with advanced vision and expert systems capabilities, a naval battle management system which would forecast likely events, suggest different courses of action, develop detailed action plans, resolve conflicts between competing goals, etc. (Schatz and Verity, 1984). During five years the program is expected to consume U.S.$ 600 million.

Some titles of projects which were already completed:

- "Decision Making in Large-Scale Military Simulation: A Requirement for Expert Systems"
- "Concepts for Army Use of Robotic - Artificial Intelligence in the 21st Century"
- "Artificial Intelligence Applied to the Command, Control, Communication, and Intelligence of the U.S. Central Command"
- "TAC II: An Expert Knowledge Based System for Tactical Decision Making"

Waterman (1986) describes 25 military expert systems which have already reached prototype stage.

There is much evidence that not only the USA invests so heavily on military applications of AI, in other countries it is just less made known to the public. In any case, an increasing considerable amount of money in the military sector is being spent on AI.

AI AS A MEANS TO REDUCE INTERNATIONAL TENSION

The enormous spending on AI in the military sector in order to improve defense raises the question if AI could not also be used to reduce international tension and to increase mutual understanding, especially in the situation of crises. Such applications of AI could eventually reduce the need to spend so heavily on armament and thus free money and efforts for peaceful purposes.

Three potential approaches are being suggested here which might fulfil these aims. Their common characteristics are that

- they shall make peaceful use of the latest AI techniques,
- they shall be jointly developed by scientists from the USA and the USSR, and
- the resulting AI programs are to be portable, i.e. it is possible to execute them on computers both in the USA and the USSR.

APPROACH 1: INTERCULTURAL KNOWLEDGE BASE

An important area, in which AI researchers have been successful, is the representation of knowledge. This representation can for instance consist of rules and meta-rules, of importance e.g. for expert systems, or it can consist in semantic nets or frames as often used in natural language systems (Kobsa, 1984).

The representation of the real world, the so-called real world knowledge base (Trost and Steinacker, 1983), is of especial importance in natural language systems. For instance, in order to understand the sentence "Sorry, I'm late 'cause I missed the bus" the system has to "know" that bus is a means of transportation, that it cannot be stopped anywhere like a taxi, that "late" means that there is a sequence of events, in which the departure of the bus is located before the arrival at the bus-stop, and so on. Parts of the whole world have thus to be mapped onto a formal structure, representing a kind of ontology (Trappl et al., 1982).

Attempts have been made to arrive at a common knowledge base for specific areas: As an example, Dr.Vadim Sadovsky of the USSR Academy of Sciences and Professor Stuart Umpleby of the George Washington University have organized a series of meetings of Soviet and American cyberneticians and systems theorists to compare and thus clarify the conceptual structures in cybernetics and general systems theory (see Part 2 of this volume). However important such efforts may be, their limitations are not only the narrow subject, the even more important drawback is the solely verbal basis. Formal representations, which demand verbal discussion as prerequisite, need to be so precise as to enable the execution of procedures on them.

The first proposed project is therefore the development of a common Intercultural Knowledge Base, developed jointly by scientists from the USA and the USSR. This knowledge base should help clarify both the differences and the views shared on important aspects of our world. The pressure to formalize the knowledge, probably in a structure combining semantic nets and frames as e.g. in KL-ONE (Brachman et al., 1978; Schmolze and Brachman, 1982) or KRYPTON (Brachman et. al., 1985) and/or in refinements of scripts (Schank, 1982) should lead to a precise AI ontology. The Intercultural Knowledge Base should thus help to improve mutual understanding.

APPROACH 2: ENGLISH-RUSSIAN/RUSSIAN-ENGLISH TRANSLATION PROGRAM

The first translation programs performed only literal translations. These led to the translation of "hydraulic ram" as "water goat" or of the phrase "out of sight, out of mind" as "invisible idiot". Present-day translation programs consider also the meaning of a sentence. Therefore, they do not only parse, i.e. grammatically analyze a sentence, but they also try to "grasp" the meaning of a sentence by mapping it onto (parts of) their world representation. Only then do they attempt to formulate the output in another language.

Several English-Russian and also Russian-English translation programs have been developed and are in use e.g. at the U.S. Air Force, at the NASA, at Yale University, at the EEC, at the University of Leningrad or at the All Union Center for Translations in Moscow. Most of these programs are doing mainly literal translations, however, some of them also "understand" what they are translating, as e.g. SAM of Yale University.

All these programs have been and are being developed in the USA by American scientists or in the USSR by Soviet scientists or in another country by the scientists of that country. None of them has been developed jointly. It is now being proposed to develop an English-Russian/Russian-English translation program by a mixed Soviet/American team which can be executed on computers both in the USA and the USSR.

Some of the benefits of this program are:

- It will enable both countries to immediately produce translations automatically authorized by the other country.
- It will make possible to check translations in advance, i.e. to see how the translation performed by the program in the other country will look like and, in the rare cases where misinterpretations might possibly occur, correct the input text accordingly.
- It will be especially useful in times of crisis when the 'hot-line' (actually a telex connection) is used.
- If sometimes in the more distant future electronic mail systems like TYMNET or TELENET (very likely not ARPANET) cross the borders, the communication between people in both countries would be facilitated by the existence of such a program.

APPROACH 3: CRISIS HANDLING EXPERT SYSTEM

Expert systems are a special type of AI programs which contain the knowledge of experts and can reason on this knowledge. The knowledge is often stored in a different way than in most real world knowledge bases (see approach 1), namely in rules and metarules. Rules are of the general structure "If (condition) then (action)". The conditions may consist of several sub-conditions which are connected by "or(s)" and/or "and(s)". As an example, a simple rule from MYCIN, a program for diagnosis and therapy selection of infectious diseases (Buchanan and Shortliffe, 1984):

If 1) the gram stain of the organism is negative,
      and
    2) the morphology of the organism is rod,
      and
    3) the aerobicity of the organism is anaerobic,

Then there is suggestive evidence (.7) that the
      identity of the organism is Bacteroides.

The rules thus contain the knowledge and also govern the way of reasoning. Rules are invoked by the "inference engine", which uses special methods, e.g. backward-chaining, to logically "connect" the rules. It would be extremely time-consuming in large expert systems to search through all rules whether a specific condition is fulfilled. Therefore meta-rules which contain knowledge about the rules guide their invocation (Hayes-Roth et al., 1982).

Expert systems consist of 5 sub-systems:

- The knowledge base,
- the inference engine,
- the user interface,
- the explanation part, and
- the knowledge acquisition component.

This last component which explains to the user how the system has come to a certain conclusion, has proved to be of special importance: It not only increases the acceptance by the users but it also helps to identify and to correct logical errors or simple bugs in the system.

There are already hundreds of expert systems in a diversity of areas, many of them are fully integrated in everyday decision-making. Expert systems have proved very useful tools by giving recommendations e.g. in the interpretation of sensor data from bore holes to advise about the chances of finding hydrocarbon, oil or gas, or where to drill the next holes (DIPMETER ADVISOR). Digital Equipment Corporation has developed an expert system (XCON) which, after the arrival of an order, configures a VAX computer system. This system has been operational since 1980 and has processed more than 20.000 unique orders with 95 to 98% accuracy since then (Kraft, 1984). DENDRAL infers the molecular structure of unknown compounds from mass spectral data, better than many human experts; MOLGEN assists the geneticist in planning gene-cloning experiments in molecular genetics; ACE identifies trouble spots in telephone networks and recommends appropriate repair and rehabilitative maintenance; etc. etc.

The development of a crisis handling expert system, in contrast to military applications already mentioned, is not aimed at winning a war but at avoiding it. Many crises led to a war, luckily not all of them. What are the conditions, what are the actions which prevent the outbreak of a war? Evidently, there are many types of conditions (military, economic, social, geographic, historic, etc.) and many types of actions, thus a first step of this project must consist in an attempt to establish a typology of conditions and actions. Surprisingly, not even the USA has an "institutional memory" for crisis management (Smith, 1984), however, crisis management is given increasing consideration (Roderick, 1985).

Although the Crisis Handling Expert System will be of enormous complexity, present-day large expert systems can already handle more than 2000 rules. The "threat assessment system CAT" (for Command Action Team), currently being developed by Carnegie-Mellon University for a new aircraft carrier of the U.S. Navy, is expected to contain even more than 10000 rules. Powerful software tools and dedicated computers (LISP-machines) have been developed to aid in the development of very complex expert systems.

While the task is still extremely difficult, the advantages of this system will be also significant:

- To identify and understand the point of view of the partner,
- to have a common basis for negotiations,
- to exchange informations rapidly in situations of crisis.

Under the hopefully valid assumption that both superpowers do not want a war and thus also would not use such a system for fraud (which is a risk), it can help to prevent the outbreak of a war and to calm down international crises.

In any case, already the joint development of the Crisis Handling Expert System would have a beneficial effect regarding mutual understanding.

PILOT STUDIES

Most likely it will be impossible to find sufficient financial support to implement all 3 approaches as projects right away. What is more, it would be most unwise. Before that the best strategy in terms of technical, political, and financial feasibility would have to be carefully considered and decided upon.

Therefore, pilot studies for the above-mentioned approaches should be carried out in order

- to obtain the principal permission of nations and institutions to co-operate,
- to contact scientists (not only AI researchers, but also linguists, translators, sociologists, historians, etc.) who might be willing to cooperate,
- to select the appropriate AI tools and techniques,
- to develop a project plan with respect to milestones, site(s), costs.

Each of these pilot studies may require 2 scientists, probably half-time working, for 1 - 2 years, plus some overhead.

CONCLUSION

It is difficult to make forecasts of the actual costs, but in any case they will be lower than those of one mid-range missile.

Some of the approaches suggested may sound naive and, perhaps, they are. But how wise is it to build an expert system for an autonomous cruise missile? In any case, from the present state-of-the-art of AI there is a chance for success. Critique, comments, offer to help or cooperation are wholeheartedly welcome.

ACKNOWLEDGEMENTS

I am grateful for ideas, comments, criticism, help already received from Prof.Bela H. Banathy, Dr.Harold Chestnut, Ms.Gerlinde Hinterleitner, Dr.Werner Horn, Dr.Nils J. Nilsson, Ms.Karin Schmid, Prof.Len R. Troncale, Dr.Harald Trost, and Ms.Christa Zeller.

REFERENCES

Barr, A., and Feigenbaum, E.A., eds., 1981, "The Handbook of Artificial Intelligence, Vol.1", HeurisTech Press, Stanford, California.

Barr, A., and Feigenbaum, E.A., eds., 1982, "The Handbook of Artificial Intelligence, Vol.2", Pitman Books, London.

Brachman, R., Ciccarelli, E., Greenfeld, N., and Yonke, M., 1978, "KL-ONE Reference Manual", Bolt Beranek and Newman Inc., BBN-Report 3848.

Brachman, R., Pigeman, V., and Levesque, H., 1985, "An Essential Hybrid Reasoning System: Knowledge and Symbol Level Account of KRYPTON", in:"Proc.9th Int.Joint Conf.Artif.Intell.", Los Angeles, CA.

Buchanan, B.G., and Shortliffe, E.H., eds., 1984, "Rule-Based Expert Systems - The MYCIN Experiments of the Stanford Programming Project", Addison-Wesley, Reading, MA.

Charniak, E., and McDermott, D., 1985, "Introduction to Artificial Intelligence", Addison-Wesley, Reading, MA.

Cohen, P.R., and Feigenbaum, E.A., eds., 1982, "The Handbook of Artificial Intelligence, Vol.3", Pitman, London.

Hayes-Roth, F., Waterman, D., and Lenat, D., eds., 1982, "Building Expert Systems", Addison-Wesley, Reading, MA.

Kobsa, A., 1984, Knowledge Representation: a Survey of its Mechanisms, a Sketch of its Semantics, Cybernetics & Systems, 15:41-90.

Kraft, A., 1984, XCON: An Expert Configuration System at Digital Equipment Corporation, in: "The AI Business", P.H. Winston and K.A. Prendergast, eds., MIT Press, Cambridge, MA.

Roderick, H., 1985, Crisis Management: Preventing Accidental War, Technology Review, 8:50-59.

Schank, R.C., 1982, "Dynamic Memory: A Theory of Reminding and Learning in Computers and People", Cambridge University Press, Cambridge, London, New York, and Sidney.

Schatz, W., and Verity, J.W., 1984, DARPA's Big Push in AI, Datamation, 2:48.

Schmolze, J.G., and Brachman, R.J. 1982, "Proceedings of the 1981 KL-ONE Workshop", Fairchild FLAIR TR-4.

Smith, E.T., 1985, A High-Tech Market that's not Feeling the Pinch, Business Week, July 1.

Smith, R.J., 1984, Crisis Management under Strain, Science, 225:907-909.

Trappl, R., Leinfellner, E., Steinacker, I., and Trost, H. 1982, Ontology and Semantics in the Computer, in: "Language and Ontology, Proc. 6th International Wittgenstein Symposium", Hoelder-Pichler-Tempsky, Vienna, Austria.

Trappl, R., 1986, Artificial Intelligence: A One-Hour Course, in: "Impacts of Artificial Intelligence", R. Trappl, ed., North-Holland, Amsterdam and New York.

Trost, H., and Steinacker, I., 1982, Representing Real World Knowledge in a Semantic Net, in: "Progress in Cybernetics and Systems Research, Vol.XI.", R. Trappl, N.V. Findler, and W. Horn, eds., Hemisphere, Washington, D.C.

Waterman, D.A., 1986, "A Guide to Expert Systems", Addison-Wesley, Reading, MA.

Winston, P.H., 1984, "Artificial Intelligence, 2nd Ed.", Addison-Wesley, Reading, MA.

STEPS IN THE CONSTRUCTION OF "OTHERS" AND "REALITY":

A STUDY IN SELF-REGULATION

Ernst von Glasersfeld

ABSTRACT

The author advocates a change of perspective concerning the concept of knowledge. He suggests that the experiential reality in which we live and in which our sciences operate is the result of a self-regulating organism's construction and should not be confounded with the ontological reality that philosophers have vainly searched for throughout the history of Western epistemology. The notion of <u>viability</u> gives a new slant to an instrumentalist theory of knowledge that serves as a basis for the cognitive construction of Others and may ultimately provide a starting-point for the development of a constructivist ethics.

INTRODUCTION

Much of what I am going to suggest in this essay goes against notions that are widely held but rarely examined for their hidden presuppositions. We were all taught that there is virtue in "objectivity" and that it is sinful to question whether objective knowledge can ever be attained. Socrates, who proclaimed that it could not, was promptly put to death. The skeptics, who ever since then have maintained the impossibility of "true objective knowledge", did not make themselves popular either. The philosophical profession, having been unable to counter their arguments satisfactorily, nowadays tends to consider them a persistent nuisance.

In this paper I shall present a few key characteristics of a theory of knowledge that is subversive in that it takes the skeptics seriously, but is also constructive in that it provides a working hypothesis that permits a non-contradictory analysis of the activity of <u>knowing</u>. Above all, I shall try to show that the experiential reality in which we live and in which our sciences operate should be considered the result of self-regulating construction and should not be confounded with the <u>ontological</u> reality which most philosophers, be they realists or idealists, bourgeois or marxist, have been searching for.

THE NOTION OF KNOWLEDGE

In the philosophical tradition of the Western world, the concept of "knowledge" has almost without exception been understood to imply that the structures that result from cognition must in some way correspond to an <u>external</u> reality; "true" knowledge was supposed to depict or replicate what is real; and "reality" was intended to refer to a world "in-and-for-itself", a world that exists ready-made, fully-structured, and independent of any cognizing subject.

At the beginning of that tradition, the skeptics had already pointed out that this conception of "knowledge" leads to a paradox. In order to test the required match between such knowledge and what it was supposed to be knowledge of, the experiencer would need some other access to the postulated reality; and that access would have to be immediate, so as to bypass the subject's activity of knowing. Within the realm of the rational, however, no such immediate access seemed to be logically possible. This impasse has not been resolved in the course of philosophical history since then. In spite of countless attempts the paradox is as solid as ever, because the established conception of what "knowledge" ought to be has remained the same throughout. There were, of course, individual dissenters, such as Montaigne, Mersenne, Vico, and a few others, who realized that knowledge did not and could not live up to the general wishful expectation. Although these thinkers made valiant efforts to revolutionize epistemology, they had little impact on the tradition.

SELF-REGULATION

Only with the advent of control theory did it become possible to conceive of models of organization and government inside an organism and to view the cognitive enterprise as an outcome of self-regulation. Insofar as these models are self-contained with regard to cognition or "information", they open a new perspective on epistemology. This possibility was slow to be realized because control theory was for the most part developed by engineers for whom the "feedback loop" was simply a powerful tool to construct highly efficient gadgets to which one could delegate certain tasks of guidance or control (Powers, 1978). These gadgets manifested goal-directed action, but the goals they pursued were of course the goals of the engineers who designed them (see Pask's distinction of purpose of and purpose for, 1969).

The actual mechanical successes during this early infancy of cybernetics did much to obscure the possibility of applying the new concepts of circular causality and equilibrium in self-contained systems to the modeling of living organisms, the phenomenon of evolutionary adaptation, and, ultimately, to the problems of cognition.

It is well to remember that Jean Piaget, the most epistemologically oriented of the modern psychologists of cognition, formulated the core of a cybernetical theory of knowledge when he wrote more than a decade before the official birth of cybernetics: "Intelligence organizes the world by organizing itself" (Piaget, 1937, p.311). This clearly and uncompromisingly created a new perspective on cognition and placed the emphasis on two hitherto neglected aspects: self-regulation and the endogenous construction of knowledge.

"FIT" INSTEAD OF "MATCH"

Once the cognizing subject is no longer seen as a passive receiver of "information", there is a radical shift of orientation. Perhaps the most dramatic consequence of that shift concerns the concept of "knowledge". Instead of the paradoxical requirement that knowledge should reflect, depict, or somehow correspond to a world as it might be without the knower, knowledge can now be seen as fitting the constraints within which the organism's living, operating, and thinking takes place. From that perspective, then, "good" knowledge is the repertoire of ways of acting and/or thinking that enable the cognizing subject to organize, to predict, and even to control the flow of experience. From this changed point of view, then, the cognitive activity does not strive to attain a veridical picture of an "objective" world (a goal which, as the skeptics have always told us, is unattainable), but it strives for <u>viable</u> solutions to whatever problems it happens to deal with.

This shift in the conception of knowledge is radical in more than one way. Not only is the notion of an absolute "truth", a truth that matches ontological reality, abandoned, but with it the notion that each problem can ultimately have only one "true" solution must be given up. Unlike the conventional concept of "truth", the concept of "viability" is not ex- clusive but reflects the common experience that the problems we face have, as a rule, more than one solution. Of course, this does not mean that all solutions to a problem must be considered equal. On the contrary, on a higher level of operating, where speed, economy, and even aesthetics are considered relevant factors, a solution may cease to be adequate, not be- cause it does not attain the goal, but because it is too slow, too costly, or too cumbersome. The main conceptual shift, however, is in relinquish- ing the idea that true knowledge should be a veridical picture of an ob- jective world.

Warren McCulloch said, in his 1948 lecture at the University of Virginia, that the break-down of a hypothesis is "the peak of knowledge" (McCulloch, 1965, p.154). It was a declaration of the break-down of traditional epistemology. When a plan of action or a conceptual structure (such as a hypothesis) fails, then and only then may we say that we have made contact with "reality" in the traditional sense. That contact, how- ever, is at best a clash between our acting or thinking and the con- straints within which our acting and thinking must take place. Such clashes may tell us something about our ways of acting and thinking, but they cannot provide a picture of the "real" world; they merely provide an indication of the insufficiency of the particular way of acting or think- ing we have embarked on. When, on the other hand, an action or conceptual structure turns out to be successful, it tells us that we have remained within the constraints, have found a way that does not clash with anything - and this, again, can provide no picture of the "real" world, because, in this case, we have done no more than act or think in the space the "real" world left unencumbered and free for us to act or think.

The fact that the concept of <u>knowing</u> was entangled with ontology and the notion of <u>being</u> from the very start of epistemological enquiry (cf. Plato's <u>Meno</u> or <u>Theaetetus</u>) has had profound consequences. One of them was to make it exceedingly difficult to expound a theory of knowledge that cuts loose from "existence" and focuses exclusively on the cognitive activity and its results. The perennial idea that the knower, even if logic shows it to be impossible, must at least <u>try</u> to discover what the world is "really" like , is an idea that is not easy to discard. Although I believe that this idea must be discarded, I shall not press the point here. Instead I would ask you to consider this: if there is to be a knowing subject who can acquire "knowledge", there must also be available some raw material, some "basic elements" out of which that knower can compose the structures which he or she is going to call "knowledge". Such raw material or basic elements are usually supposed to be "data" or "in- formation" that is conveyed to the knower from the "outside" via the senses. If, for the moment, we accept that realist hypothesis, it does not alter the fact that it is still the knowing subject who has the task of <u>composing</u> the data or <u>interpreting</u> the information in order to achieve a "representation" of reality. Whichever way one looks at it, therefore, "knowledge" cannot be a commodity that is found ready-made but must be the result of a cognizing subject's construction. It is this constructive activity that we shall now look at a little more closely.

THE INITIAL CONSTRUCTION OF REALITY

What we ordinarily call "reality" is, of course, the reality of the phenomenal, the reality of the relatively durable perceptual and conceptual structures which we manage to establish, use, and maintain in the flow of our actual experience. This experiential reality, no matter what epistemology we want to adopt, does not come to us in one piece. We build it up bit by bit, and the construction is achieved by a succession of steps that come to form a succession of levels.

Repetition is an indispensable factor in that development. A simple sensory impression, a flash of color, for example, remains a dubious experience if we are unable to make it recur. Similarly, our concept of "existence" (in the sense of an experiential item "being there", in its own right and independent of our experiencing it) is indissolubly tied to a notion of "permanence"; and permanence, after all, can be conceived only on the basis of at least two moments of experience, two moments that can be linked to constitute a continuity or to frame an interval during which the experienced item could be said to perdure. Such links have to be made - and, in order to be known, they have to be made by the knower.

To a realist, it may sound absurd to say that repetition or recurrence has to be <u>constructed</u> by the experiencing subject. To recognize something as having occurred before is so commonplace, so "natural" an experience that it is easy to ignore (though the word clearly indicates it) that recognition is a cognitive <u>activity</u> and, as such, requires some doing. In this context it is essential to remember that I am speaking of epistemology, i.e. of <u>knowing</u> and not of <u>being</u>. In order to conclude that A is the same as B, a comparison must be made and this comparison must yield the result of "sameness" rather than "difference"; and comparisons do not make themselves, they have to be made by an active agent.

The notion of "sameness", without which we could not know that A is a repetition of B, is itself a much more complicated affair that it seems at first sight. Apart from the fact that it is always the cognizing subject who chooses the property or dimension in which two items are considered "the same", the very concept of "sameness" involves an ambiguity that is of fundamental epistemological importance.

An example may help to make this ambiguity transparent. From the two statements "Julia bought the same dress as Ann" and "Julia slept in the same room as Ann" we would normally infer that, respectively, they imply two dresses but only one room. Of the two dresses we are told that they are equivalent in all respects one usually considers when comparing dresses; of the room, on the other hand, we assume that it is one and the same. The expression "the same", thus, leads us to construct a relation of <u>equivalence</u> in the one case, and a relation of <u>individual identity</u>, in the other. Conceptually, the two relations are quite different: the one constitutes the basis for the formation of classes, i.e., collections of items that are considered "the same" in some respect; the other constitutes the basis for the construction of what Piaget has so aptly called "object permanence", i.e. the notion that things have a life of their own and <u>exist</u> even during those intervals when they are not within the subject's immediate experiential field. For a fuller exposition of this conceptual analysis, see Glasersfeld, 1984.

To continue with the "levels of reality", a somewhat higher level is achieved whenever we are able to coordinate an experience in one sensory mode with an experience in another sensory mode. If an item (e.g. a patch of color) which we have isolated in our visual field can be recurrently "corroborated" by tactual exploration (e.g. a palpable edge), or when its appearance or disappearance can be coordinated with an auditory experience, then that item will be considered a good deal more "real" than if it remained exclusively visual.

There are many shades and subtle degrees in this construction of reality. The stars became more real once we were able to plot their motion, and the moon became more real to the man who stepped on it. But whatever reality each one of us creates for him- or herself is still precarious, because we all may have experiences that contradict our ordinary construction. We are subject to sensory illusions, have dreams, and sometimes experience hallucinations. At times, these irregularities cast doubts upon the reliability of our senses.

Fortunately we have another, much more powerful method of confirming the reality of our experiential world: the corroboration by Others. The way this method works, however, is, again, a great deal more complicated than it may seem at first sight. If I were to ask my neighbor for a carafe of water on the table in front of us, and he passed it to me, this would at once allay whatever doubts I might have had about the reality of the carafe. Things seen by several observers are taken to be more reliably "real" than things seen by only one. Very often, however, this corroboration by Others is taken to imply much more than simple experiential compatibility. Indeed, it is usually assumed that if you and I agree that we perceive, say, this carafe, our agreement could demonstrate not only that we have compatible experiences but also that what we experience must be a "true" reflection of what exists in an independent reality, whether we experience it or not.

In other words, the probability that Others experience something that seems compatible with what we ourselves experience, is usually taken as evidence that the "shared" experience reflects an independent "objective" reality. But David Hume, more than two centuries ago, showed quite conclusively that such inferences are based on faith, not on logical necessity.

Now, however, I would claim that these inferences or assumptions are illusory on two further counts, because both the process of communication and the Others with whom one communicates are not as straightforward as they may appear. In the two sections that shall follow I shall argue that neither communication nor the Others with whom we populate our experiential world can bring us closer to knowing an ontological reality.

UNDERSTANDING OF CONCEPTUAL FIT

Though corroboration by Others can be obtained without speaking (in the above example, I might simply have pointed at the carafe and the neighbor might have passed it wordlessly), in the overwhelming majority of cases corroboration is obtained by means of linguistic interaction. Language, thus, becomes an almost indispensable instrument in the construction of a "shared" reality. But here, once more, a confusion concerning the use of the expression "the same" creates the illusory impression that the shared reality "exists" and, therefore, must be independent of the communicating subjects who happen to share some knowledge of it.

To show that this _is_ an illusory notion, we have to look more closely at how we come to "understand" language, communicatory gestures, or any other conventional semiotic system. Among other things, semiotic systems involve the formation of associations which link certain auditory or visual percepts (signals, signs, words, etc.) with specific _other_ segments of our individual experience. The particular "meaning" of these communicative items is established gradually through interaction, in that the segments of experience a subject has associated with a particular sign or word are modified and adjusted until they _fit_ into the situational contexts of recurrent communication events. When we say that we understand a piece of language, we are in fact saying that we are able to fit our _interpretation_ of that piece of language into our assessment of the situation in which it was uttered, as well as into such as we might make of the speaker's or writer's intentions. All of this is and remains part of our _own_ experiential world, the world that we have conceptualized during our past experience, and does not and cannot attain at any point a "real" world that is supposed to be independent of the language-user's conceptualizations.

Language, thus, can function quite as well as it does, _without_ "reference" in the philosopher's sense, i.e. without referring to "objective" entities or events outside the experiential worlds of the members of a linguistic community. Hence, the fact that my neighbor "understands" what I want when I ask him for the carafe of water merely requires that the subject-ive experience my neighbor calls "carafe" is sufficiently like the subjective experience I myself have associated with that word. This does _not_ entail that ontological reality contains a replica of either his or my carafe experience - it merely presupposes a "reality" that is rich and ample enough so that both he and I can construct such experiences in the given context.

In other words, there is no reason to assume that his carafe experience and mine are "the same" - all that is needed for my request to be understood, is that what he constructs as "carafe" satisfies the constraints I myself had in mind when I used the word. And if no match but only _fit_ is required between his experience and mine, it would be downright absurd to assume that there should be a match between both our experiences, on the one hand, and an object in ontological reality, on the other.

OTHERS AND THE NOTION OF OBJECTIVITY

This second illusory assumption springs from a habit of thought that is so deeply ingrained that we tend to take it as an "obvious" common sense fact: we consider our construct of Others to be unquestionable and _ontologically real_. Yet, how could it be? Our knowledge of Others, like all the furniture of our experiential world, must originally have been composed out of elements of our own experience. I have elsewhere tried to give an account of how a cognizing organism may come to attribute the capabilities of perception, representation, cognition, and goal-directed action to certain items in the experiential field (von Glasersfeld, 1981, 1982). Once they have been made, these attributions may seem to be based on (and therefore confirm) a mystical belief or metaphysical conviction. In a rational model of the cognizing subject, however, neither of these sources would be admissible, because, by definition, they preclude any further rational analysis. If we want to model the knowing subject, we should avoid all presupposition of ready-made, intrinsic knowledge.

The notion of Others - in the sense of particular items to whom one concedes capabilities similar to those that one attributes to oneself - this notion can be grounded in the relatively simple realization that it may be advantageous to create a class of rather special items in one's experiential field, namely items whose actions seem easier to predict and to control if one _does_ assume that they have some of these prized capabilities. A butterfly perched on a flower in the field, for instance, will be easier to catch if one hypothesizes that it, too, _can_ see and, therefore, will react to quick movements or changes of light and shadow in its immediate environment. Similarly, if I speak in order to induce my neighbor to pass the carafe of water, I must hypothesize that the speech sounds I emit will be perceived by him and will trigger certain cognitive operations and, eventually, motor acts that are somewhat compatible with those someone else's utterance of these speech sounds would trigger in me. Hypotheses of that kind have worked fairly well in my past experience, and thus I have come to construct Others more or less in the image of myself.

This is by no means a new idea. Kant, in the first edition of his _Critique of Pure Reason_ (1781), has a passage that expresses it with perfect clarity (p.223):

> It is manifest that, if one wants to imagine a thinking being, one would have to put oneself in its place and to impute one's own subject to the object that one intended to consider ...

(Remark: I am indebted to Eduard Marbach for having made me aware of this seminal passage in Kant's early work. The translation is mine.)

What I find particularly seductive about this idea of imputing the model of oneself to Others, is that it provides a starting-point for an analysis of the concept of "objectivity". From the constructivist perspective, as I explained earlier, "knowledge" comprises those constructs which the acting and knowing subject finds useful or at least tenable in the face of further experience. Clearly, then, if some of these constructs turn our to be _viable_ not only in one's own organization of experience but also as hypothetical basis for the computations and calculations one imputes to Others whose actions one wants to explain, predict, and control - well, then one will almost inevitably come to think that these constructs are less "subjective" than those which one can find instantiated only in one's own operating. As a constructivist, then, I can agree that knowledge should be called "objective" if it serves not only me, the subject, but also my interpretations of Others and _their_ actions and understandings. But this does not and cannot warrant the assumption that, because a cognitive structure turns out to be useful in the interpretation of Others, it must on that account reflect, depict, or convey anything about the structure of an ontological reality.

A CONSTRUCTIVIST APPROACH TO ETHICS

The introduction of the concept of Others and, with it, of the level of experiential reality on which "objectivity" becomes operative, must sooner or later raise the question of ethics. At first it might, indeed, seem that _no_ ethics whatsoever could be founded on an epistemology that is so explicitly centered on the cognizing subject. But this, I believe is not necessarily the case. Let me say at once that it is only recently that I have begun to think about this problem and what I have to offer is at best the suggestion of a starting-point.

The main difficulty in attempts to provide a rational grounding for ethics has always been that, whatever rules of conduct one wanted to justify, their justification required the assumption and acceptance of certain values - and the scale on which these values were to be assessed invariably turned out to be itself in need of justification. In the con-

structivist epistemology, I believe, it can be shown that at least a rudimentary ethics can be <u>logically</u> developed out of the very same assumptions
that underly the central notion of the construction of knowledge.

This development can be summarized in the following steps:

(1) The model of the cognizing organism involves the working hypothesis that the organism's intelligence is essentially self-regulating and aims at coordinating its experiences in such a way as to
render experience explicable and manageable.

(2) Insofar as these coordinations or constructions are successful or
<u>viable</u>, they constitute the organism's experiential reality.

(3) Since criteria of viability vary with the context, there are
levels of reality (repetition, mutual corroboration of sensory
modes, confirmation by Others).

(4) The constructs that constitute this subjective reality will be
considered "objective" if they turn out to be viable also in the
construction of the cognizing subject's models of Others (i.e.
other cognizing subjects).

In order to achieve the highest level of "reality", therefore, the
cognizing subject not only <u>needs</u> Others but must also construe these
Others with concepts that are not incompatible with those used in the construction of him- or herself; and, in order for these concepts to be and
to remain viable not only for oneself but also for Others, one must necessarily assume that these Others operate within a goal structure that <u>could
conceivably</u> be one's own. It thus becomes clear that what Kant proposed
as his Categorical Imperative is not merely an ethical prescription but
is, in fact, a requirement of the individual's own construction of a viable "objective" reality.

Both formulations of the Categorical Imperative are equally pertinent
(Kant, 1788). "Act always in such a way that the guideline of your action
could be taken as guideline by all Others", means precisely that the
models you construct of Others, in order to serve you as corroboration of
your own reality, must be potential analogs of yourself, at least with
regard to their goals and to the operations used to attain them. Thus, if
the model you construct of yourself is to remain a viable one, the goals
and operations you choose to construct it must always remain compatible
with those which turn out to be viable in the construction of your models
of Others.

As for Kant's other formulation, "Treat Others as ends in themselves
rather than as means to your own ends," yields the interpretation that, in
order to establish the highest level of your own reality, you must concede
to Others the need and the possibility to construct their own reality.

The tentative claim I should like to make is clearly not that here is
the possibility of an ethics that can do <u>without assuming any a priori
value</u>. Rather, I would make the much more modest claim that constructivism may point the way to developing an ethics that requires no <u>further
assumptions</u> than those that are inherent in the constructivist theory of
knowledge.

CONCLUSION

It has taken many years to clarify the psychological feasibility of the processes of conceptual construction that are indispensable for a constructivist theory of knowledge. That such a clarification was possible at all is due to the pioneering work of Jean Piaget, who was the first to attempt an <u>operational analysis</u> of the mind. Given the revolutionary direction of the step he took in positing the human mind as a mechanism of self-regulation, it is of relatively little importance whether or not the actual operational analyses he produced can, in every detail, attain "objectivity" in the sense I have suggested above. Even if his contribution should eventually be reduced to the mere launching of a cognitive psychology that treats the organism as an informationally closed system that works to maintain its equilibrium in the flow of experiential perturbations, it would still have the enormous merit of having introduced a new and promising basis for the development of a non-contradictory theory of knowledge.

In order to survive, the constructivist theory of knowledge must be able to withstand attack. The present task, therefore, is to devise a reasonable defense against the philosophers' almost immediate objection that any such constructivism is merely a new mask for the spectre of solipsism. This defense hinges on the change in the conception of "knowledge". The notion of cognitive constructs that remain <u>viable</u> in the face of further experience, leads to a conception of "knowledge" which, though subject-generated, cannot be brushed aside as idealism because it <u>does</u> take into account the existence of an ontological reality. This ontological reality, however, is no longer a reality to be <u>known</u>, but rather a reality that constrains the range and the success of all acting and cognitive constructing.

This peculiar <u>negative</u> relation of knowledge to reality should not come as a great surprise to cyberneticians. Any self-regulating device, after all, "knows" only what it senses, and "acts" only when what it senses does not fit or satisfy the conditions or patterns that have been chosen as reference. In any cybernetic gadget, therefore, equilibrium is achieved whenever an interpretation of the sensory signals conforms to a pre-established desired pattern. Similarly, in the cognitive subject, equilibrium is achieved whenever the experiential situation can be satisfactorily managed without reorganization of the relevant conceptual structures.

The constructivist theory of knowledge is explicitly and unashamedly "instrumentalist". Yet it should be immune to the usual arguments that attack instrumentalism because of its traditional connection with utilitarianism. For the constructivist, knowledge is not an instrument in the struggle for material benefits in an independent "objective" reality, but an instrument of equilibration in the cognizing subject's experiential world.

REFERENCES

Kant, I., 1781, Kritik der reinen Vernunft, 1. Auflage, (Gesammelte Schriften, Bd.IV). Koenigl. Preussische Akademie, 1910ff., Berlin.

Kant, I., 1788, Kritik der praktischen Vernunft, (Gesammelte Schriften, Bd.V). Koenigl. Preussische Akademie, 1910ff., Berlin.

McCulloch, W.S., 1965, Through the den of the metaphysician, <u>in</u>: "Embodiments of Mind", MIT Press, Cambridge, Mass.

Pask, G., 1969, The meaning of cybernetics in the behavioral sciences, <u>in</u>: "Progress of Cybernetics", J.Rose, ed., Gordon and Breach, New York.

Piaget, J., 1937, "La construction du reel chez l'enfant", Delachaux et Niestle, Neuchatel.

Power, W.T., 1978, Quantitative analysis of purposive systems: Some spadework foundations of scientific psychology, <u>Psychol.Rev.</u>, 85:417.

von Glasersfeld, E., 1981, Einfuehrung in den radikalen Konstruktivismus, <u>in</u>: "Die erfundene Wirklichkeit", P. Watzlawick, ed., Piper, Munich.

von Glasersfeld, E., 1982, An interpretation of Piaget's constructivism. <u>Revue Internationale de Philosophie</u>, 36:612.

von Glasersfeld, E., 1984 Thoughts about space, time, and the concept of identity, <u>in</u>: "Book-Conference. First Instalment", A. Pedretti, ed., Princelet Editions, London/Zurich.

STEPS TO A CYBERNETICS OF AUTONOMY

Francisco Varela

In the context of this panel on some fundamental guidelines for cybernetics and systems theory I would like to present today a very specific point of view: that of a biologist. I came upon these issues because I was involved in studying things like nervous systems or immune systems. In cybernetics and systems theory, important notions have been shaped by empirical research dealing with complex biological systems. It is well known to all of you that in the pioneering days - in the 40ies and the 50ies - biology played a fundamental role in asking the questions that then led to the full development of these disciplines.

At the risk of 'dulling' something that is more interesting - let me focus on the contrast between two giants of that time: John von Neumann and Norbert Wiener. Both of them strongly motivated from biology, both of them making enormous contributions to cybernetics and systems theory, but both of them, in their latter days, moving in very different directions (Heims, 1982). Norbert Wiener by emphasizing the quality of independence, autonomy, creativity, the quality of living beings to create their meaning, to create their world. John von Neumann by emphasizing the quality of specifying decision rules, procedures for exact computations, control. During those early days, it was unclear what was going to be the dominant trend of those two sides of the issue - whether control, autonomy, or both. It seems to me that it is quite clear - looking back from the 1980ies - that von Neumann actually prevailed. Cybernetics and systems theory developed most of their effort into the now familiar characterization of a machine (or an automaton) as an input-output device: you have inputs, transformations, and some kind of output, which, of course, can be made very precise with the notion of a Turing machine. The fundamental notion of a Turing machine and all of its abstract and concrete applications is, as far as I can see it, the ultimate evolution of the notion of Cartesian causality.

The point I would like to raise is that it is high time now to develop the Wienerian side quite a lot more, and that it is only in recent years that, out of biology itself, have come the actual grounds to demand the revision of this trend in cybernetics and systems theory. The grounds for that are mostly out of the study of systems like the immune system, the nervous system, or ecological systems. So, as a biologist, my feeling - or my judgement if you want - is that cybernetics and systems theory as it is developed today is <u>incomplete</u>. It is incapable to actually encompass the totality of the relevant biological phenomenology. It is one-sided. What it leaves aside is what I referred to before as the quality of <u>autonomy</u>, the quality of living systems of having an assertion of their internal coherences, their internal determination, as well as the fact that it is this internal determination the one that shapes or imbues a world with meaning.

I am talking then about two basic issues:

(1) self-determination and
(2) the emergence of meaning.

These two qualities are completely outside, cannot be fit into the model or mechanism captured by a Turing machine. I do not mean to say that nobody has ever asked about these qualities before. The incompleteness vis-a-vis this living quality of autonomy has been raised in many different ways. For example it has been said that we must study more the question of self-organization; some people prefer to talk about synergetics; some others speak about co-operative properties. But it seems necessary that we pick up the basic issue out of these various trends in various disciplines, in order to make a more coherent picture of what this line of development of cybernetics ought to be or could be. So let me outline for you, in the next minutes, what I see as the fundamental points that we need to grapple with and deepen in our understanding, if we are to come to grips with the autonomous side of living systems. I said I am speaking as a biologist - in social systems very similar issues arise, but I am very ignorant about them; I'll speak just about those I can put my hands in.

First of all, there is the question of how to characterize a system. In the classical Turing-von Neumann context, a characterization of a system is given by the list of inputs and outputs and their transfer functions. That is to say, one characterizes a system by the way it handles what is given to it as a specified input. This is very familiar to everyone of us. For an autonomous system the characterization is different. One shifts from the emphasis on the inputs - and how they are transformed - to the emphasis on the internal regularities of how the system is constructed. I call this 'operational closure' (Varela, 1979): in an autonomous system we find that its components are so strongly interrelated that it is this internal coherence and interrelatedness what is central, rather than the way inputs are specified. So, instead of inputs and their transformation, one shifts to operational closure, as a characterization of the internal network.

What becomes of inputs then? We have shifted our emphasis: they are not something that is explicitly given, but inputs become simply a background of perturbations which are undefined, 'background noise'. They do not enter into the definition of the machine, system or procedure. What we have learned from the study of systems which have a very clear operational closure, (such as cells, nervous systems or immune systems) can be stated as an empirical conclusion that does not yet have a fundamental theoretical validation: every time there is operational closure, there is also the emergence of internal regularities which arise out of the inter-connectedness. Such internal states can be thought of as 'stabilities' or, more appropriately, one can talk about eigen-behaviors, that is, self-determined behaviors.

The study of eigen-behaviors is a big chapter which is of great interest to me, but we have to leave it at that for the time being. Instead, let me backtrack: we have proposed a shift in the characterization from a Turing to an autonomous machine; we have proposed a shift in the characterization from an input and transfer function to one of operational closure and the understanding of how that closure gives rise to eigen-behaviors. These are two fundamentally different modes of approaching and studying a specific situation.

Now, once we have this alternative characterization for an autonomous system, it immediately follows that the mode of relationship of such a system to its environment is completely different. In the first case, (a Turing automaton with inputs) the mode of relationship with its environment is always one of representation, namely items of the environment become instructions that act on the structure of a system - that's why we call them inputs, otherwise we would not call them that. It is fundamentally an instructive mode of relationship or interaction. Instead for an autonomous machine characterized by its closure and its eigen-behavior, what happens is that these eigen-behaviors will specify out of the noise what of that noise is of relevance. So, what you have is a laying down of a world, a laying down of a relevant 'Umwelt'. A world becomes specified or endowed with meaning; out of eigen-behaviors, there arises possibility of generating 'sense'. So what we are talking about here is the contrast between an instructive Turing automaton and an autonomous machine capable of creating (or generating) sense.

So far we have examined what an autonomous system could be by giving a characterization for it, by seeing its mode of relationsship with its environment. A third aspect to consider - to me the most poignant one - is that in this approach we have changed our mode of inference. We have, in fact, an entirely different mode of inference. The mode of inference in the context of a Turing automaton is one where the outside is causal to the inside, and therefore where objectivity is underlined. It is the structure of the environment that has to be well-defined and predetermined. From the point of view of the characterization of an autonomous mechanism, what happens is that the inside is what endows meaning, creates sense - therefore the mode of inference is from the inside to the outside and thereby objectivity is immediately bracketed. That is to say, we suspend what from our point of view looks like a very structured and defined world, and we let the system reveal what is relevant for it. So, objectivity from that point of view is bracketed.

Let me briefly repeat those three points:

(1) Characterization: going from input and output and transfer-functions to operational closure and eigen-behavior;
(2) Mode of relationship: going from an instructive one to laying down a sense;
(3) Mode of inference: going from underlying objectivity to bracketing objectivity.

Please do not take me to be saying that one is better than the other. I am saying they are different. I am saying we need the second, we need the characterization of autonomous mechanisms to actually come to grips with that which is presented in the living world. Of course there are some contexts, in which the Turing characterization is very good and very useful, but I am saying that this is far from being all we need. We need to develop the other tools. We could go here into a long list of the partial tools that have already been developed, the applications in many fields. I see here today some distinguished people who have contributed enormously to do this in different areas - but we don't have time to do that. Instead, let me give you an example which I hope I'll manage to convey the flavor of what I want to say.

Several years ago, a few immunologists in Europe and the United Stated discovered something that was was contrary to what was the belief up until then about the so-called antibodies. These are protein molecules circulating in the blood which normally are said to bind to antigenes; that is, molecules that come from the outside such as viruses and bacteria. Antibodies were taken to be means to guard your body by reacting to an antigene that would come from the outside, bind to the antigene, and then reduce it to nothing. So, it was like a surveillance system. Typically an input - output situation. Now, as it turns out, somebody (almost by accident) discovered that there are antibodies to other antibodies, the so-called anti-idiotypic antibodies. Idiotypes are molecular determinants in the lymphocite cells which make up the immune systems. One can have antibodies against antibodies which of course produce antibodies against those antibodies etc., etc.. If you have ever seen such an infinitely branching structure, you know that it is equivalent to a closed network. So, there is a fundamental immune closure: a lymphocyte talks mainly to lymphocytes; it is not looking outside to antigenes, but is mostly talking to 'his own peers', so to speak. What becomes of an antigene then? An antigene becomes something very different - it is that molecule which resembles enough one of these idiotypes to be able to sneak in into the ongoing closure and produce a change in the network. Antigenes are not determined as a list of what are the relevant bacteria to be kept outside, but rather by the structure of the immune system itself. The rest is simply nonsense. Thus the immune system endows the molecular world with a meaning, in the sense that it is only through its closure that certain molecular items are classified as being relevant. Furthermore, during the development of an organism, the idiotopes are never the same, so that if I were to take the response of my immune system and anyone of yours to the same molecule, we would find that both styles of response are completely different. The family of antibodies against those idiotopes will be completely different, although from the behavioral point of view both you and I would have performed a 'recognition' out of this molecule. The quality of laying down a sense is not given by what you give, but by the structure of the system. It can be arrived at by many possible rules. To say it metaphorically: you can lay down a path by walking in many different ways - the important thing is that you keep walking. That is what the immune system does - it keeps walking. You walk differently than I do, we both walk - that is what matters. There is no 'representation' of the world and its invasive agents.

The examples could be multiplied and discussed in detail, such as for the nervous system. Other people have worked on social applications of similar notions. Let me conclude. My main point today is that I am convinced that we need to understand that autonomous mechanisms are fundamentally different from Turing-Cartesian ones. They address themselves to different issues, they require different tools, they entail a different form of looking at the world. But we need to develop them in order to actually cope with situations which are of a complexity appropriate to the living and the social world. To me, what it comes down to as a fundamental guideline for cybernetics and systems research is the need to actually go back to a Wienerian spirit, and re-take the issues that he raised in the latter part of his life again afresh with new tools and with twenty years of having developed a cybernetics of Turing mechanism. Thank you.

DISCUSSION

Moderator: Any questions to Francis Varela?

From the floor: Yes, one small, and perhaps an important one: I am wondering what the reason was that referring to the founding fathers of our systems theory you have limited yourself to Wiener and von Neumann.

Varela: Just for the sake of making a point; these two gentlemen go so parallel, they are both mathematicians, they are both professors at American universities, they both dealt with war issues during the war, they both exchanged views quite a lot in their lives. It is just a good example, it is a story well told - se non e vero, e ben trovato.

From the floor: ...., if you look at the main part of systems theory you will see of course the type of systems which I would refer to dynamical systems with input and output - I am not sure if the Turing machine is really a very good example for what you mean, because I consider most applications as rather autonomous - you load a program and then you make it run, so at the initial stage ....

Varela: We are not talking about the same thing....

From the floor: I am saying, in engineering of course you are not so much concerned about autonomous systems ....

Varela: I know, that's fine ....

From the floor: The systems theory, or the biologist ... should just come with the problems and then it will be just built out in a way as the contributions ... from engineering.

Varela: That is right, again - please do not misunderstand me. I am not arguing for one type of approach (such as autonomous mechanisms) in favour of another. I am saying: this is fine and good and for engineering it works wonders, but let us not be naive that when we try to transpose that to another level of problems, they just do not work. Basically they do not work, or they work to such a limited extent that we leave out all the real meat of what the issues are. So let us just enlarge the scope. I am not saying that this is all wrong and has to be thrown out the window - no - it is another topic altogether.

From the floor: Also the word 'autonomous' is of course already very much used because we have an autonomous differentiation ....

Varela: I know, but this is in a different sense ...., in a fifteen minutes' talk the words will get fuzzy, but we should pursue that later. Please, Professor Beer ....

Beer: I've realized that the distinction you have drawn is an expository device. But I would like to know to what extent you think that W.Power is a bridge between these two paradigms? You know the work I am referring to?

Varela: Yes, 'Behavior as the Control of Perception'.

Beer: The perception is governed by the process of seeing. It seems to me that that makes a kind of bridge between the two things you have distinguished between.

Varela: No, I do not think so because it seems to me that what he is doing is applying the tool of one field to address the issues of the other. I would classify Powers definitely as a person interested in autonomy. But once you confront a system, you have to make a choice that, as far as I can tell, is always all or none. I either take this perspective, or I take this other one. I am not saying you cannot take both, but at every one moment it is either this or that. Maybe in some future somebody will find a clever way to actually finding some solid complementarity. It is a stance that we take; it is that or that, but not both of them simultaneously. Powers, I believe, is slightly misleading in that he takes one stance and then he talks with the language of the other. But that does not mean that it is a bridge.

Beer: That is what I meant.

From the floor: My question now is: What is your concept of information you have in your framework - is it information as to how to construct a system or how to alter a system?

Varela: The shortest form I have to answer that is saying that information exists on the side of what I am calling here input/output systems - we all know what it is. Here I say that this notion becomes fundamentally different; that is why I do not like to even use the word - but if I were to use it, I would write it something like this: 'in-formation' - that is, something that is formed within.

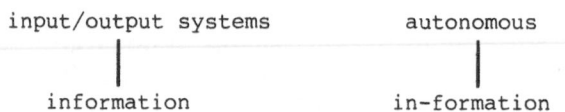

input/output systems   autonomous

information   in-formation

In other words, it is the quality of a system to endow the world with a certain regular relationship which then tells you that that item has meaning, e.g. that I am allergic to pollen. You see the difference? These two ideas have nothing to do with each other - that is why I rather drop this word and say that in this side there is no information. There is sense, there is meaning if you want.

From the floor: You said in the beginning, I think, that there are two possible ways in which this subject could go; von Neumann's way or Norbert Wiener's way. One was to do with autonomy and one with control. I was astonished to hear you say that von Neumann was to do with control - I would have thought he was using his logic instrumentally and to do it with autonomous machines and that Norbert Wiener in his book 'The Human Use of Human Beings' at least was trying to get us to trying to control how we think ....

Varela: We have a fundamentally different reading of the same people - I take it exactly the opposite. Von Neumann was the one who advocated the development and use of the atomic bomb and Wiener was the one who always stood against it. Wiener was the one who always advocated the use of technology for social and political problems. Von Neumann was the one who was always trying to make it technological and precise and computable. There is historical evidence for that (Heims, 1982). So to me von Neumann - the old von Neumann - is really leaning towards the control ideology, while Wiener more and more reacted against it and claimed a revival of the revision of values and a non-separation between value and research. I am not a historian of science, my talk is not founded on that, it was meant as an expository device. What I do want to address is the importance of the difference between the two types of mechanism.

REFERENCES

Heims, S., 1982, "John von Neumann and Norbert Wiener", MIT Press, Cambridge, Mass.

Varela, F.J., 1979, "Principles of Biological Autonomy", Elsevier/North-Holland, New York.

SECOND ORDER CYBERNETICS IN THE SOVIET UNION AND THE WEST

Vladimir A. Lefebvre

1. SECOND ORDER CYBERNETICS

In the early 1960's, cybernetics underwent surprising changes: an investigator studying Universum suddenly turned into the object of investigation. The languages of systems representation and cognitive research procedures themselves became the objects of investigation - just as morphological and functional structures were before. The process of "self-objectification" began independently in the Soviet Union and in the West. This shows that cybernetics develops according to its own immanent logic and independent of current fashions, the individual priorities of particular scientists or cultural stereotypes. On the other hand, the differences between Soviet and Western approaches make it very interesting to compare them. Their integration will allow us to see more clearly the general structure of the set of problems, methods and schemes which is called cybernetics.

The concept of "self-objectification" was in the air in Moscow's philosophical and cybernetic seminars in the early 1960's. It seems that I was the first to formulate this idea distinctly (Lefebvre, 1965). At that time I worked in a classified military institute studying the problem of how to automatize decision making; the problem of "self-objectification" appeared to me not only in its abstract philosophical form, but also as a problem related to describing the interaction of military systems. I have worked out a special formalism called "reflexive analysis" and introduced a concept called "reflexive control", useful for studying the informational influence on a system that is capable of "awareness" of itself and of the influencing system.

In 1967, I succeeded in publishing a non-classified book, Conflicting Structures, describing the main results of my work. I will take the liberty of citing two passages from this book.

> We isolate a special class of objects which we refer to as "objects comparable to the investigator in their complexity." Consider, for example, an army commander as an investigator who wishes to analyze the object in front of him - an enemy's army. He may analyze this as he would any other innocuous, ordinary looking object: by constructing a configurator consisting of two projections - spatial location of the enemy's army and its functional structure. But this analysis is not adequate to deal with the problem at hand. The commander's most important objective, from his point of view, is to discover the enemy's plans, to find out to what extent the spatial and functional structures are "natural", and to what extent they are deliberately contrived by the enemy in order for the commander to

discover and be deceived by them. In this case, the investigator has to reflect the "inner world" of the object. He needs special tools to do so; we will call these tools "reflexive" ones. The distinction between investigator and object, which is usually quite clear, now disappears. An external observer identifying himself with the investigator finds himself in a tight corner. What can he do if the object is also an investigator? The observer may assume a "pathological" position: to look at everything from the object's point of view (to analyze the investigator from the point of view of an object!) (Lefebvre, 1967, p.9-10).

In addition:

A complex organism appears as a special symbiosis of disparate structures in the same material. Several functional structures exist in one "morphological body" and each of them lives its own life ...

We illustrate this idea with an example commonly found in popular psychology books. Figure 1 contains two pictures made from the same lines: on the one hand, it is a profile of a man, on the other, a mouse. We may read this drawing in two ways, and what we see depends on our schematization.

Fig. 1.

Now let the reader imagine that the mouse and the profile live their lives independently. Let them (not an external observer) look at themselves, "feel" their entity, and try to change their shape. While waving its tail, the mouse wrinkles the neck of the profile. In order to exist, the mouse and the profile have to maintain certain obligations to each other. It is also possible that one of them could change and keep his own essential features but destroy those of the other one.

In this example, the "external observer" is identified with an object. We have created an abstract object in which several different "investigators-constructors" are made in the same material. The process of "observing" the object is closed to the object itself (Lefebvre, 1967, p.17-18).

This idea of the object-investigator is analogous to Heinz von Foerster's aphorism given in the Foreword to the collection of his works by Francisco Varela:

First order cybernetics: the cybernetics of observed systems. Second order cybernetics: the cybernetics of observing systems (von Foerster, 1981, p.xvi).

I will use the term "second order cybernetics" in this very way: it is a set of concepts and methods for the investigation of "observing systems". The ideas of second order cybernetics played a completely different role in the Soviet Union than in the West. In the Soviet Union it formed the conceptual basis for methods of influencing the enemy's decision making process during a military conflict. The concept of "reflexive control" was used for the description and planning of such an influence (see, for example, the books by V.V.Druzhinin, Deputy Chief of General Headquarters, 1972, 1976, 1982). In the West, however, the study of second order cybernetics was confined to a very small group of researchers shaping its studies in a very esoteric manner.

Western second order cybernetics developed a more elaborate epistemology, while the Soviets surpassed the West in clear formulations and the ability to solve specific problems.

2. SELF-REFERENCE AND SELF-REFLEXION

The most significant distinction between Western and Soviet approaches was determined by the West's preoccupation with biological problems and the Soviet's intense interest in psychological problems. In studying the reproductive mechanisms of biological systems, one of the main problems is how to explain the existence of an absolutely adequate "image of itself" inside the system, and how to avoid logical paradoxes of self-reference. In describing psychological reflexion, the problem of adequacy between the image of the original and the original does not arise. The main problem is to find methods of registering the differences between the image and the original. Therefore, for a biologist, the problem of self-representation is connected with classical problems of self-reference, while for a psychologist it is connected with classical problems of the inadequacy of reflexion. This led to different intentions in constructing formalisms:

West	Soviet Union
Adequate self-representation is postulated; laws of self-representation are deduced.	Laws of self-representation are postulated; all possible morphisms are considered; limitations for self-representation are deduced.

3. FORMALISMS

In Western cybernetics, the most well known attempt to construct a formal calculation for the description of self-reference belongs to Francisco Varela (1975), who modernized Spencer Brown's "Laws of Forms" 1969). The system of axioms was chosen in such a way that self-reference would exist. The Universum hidden behind this system of axioms is a sheet of paper on which an individual-operator-operand, , is living and capable of existing in copies and forming various flat configurations. The rules for "equivalent" transformations are given in such a way that the "equality" which is interpreted as the "realization" of self-reference is achieved.

My representation of systems with reflexion was given in the book Conflicting Structures (Lefebvre, 1967, 1973). Briefly, the formalization is as follows. Symbols T, x, y, z, +, (, and) are introduced. T stands for "reality", x, y, z, are interacting individuals; + is a symbol for the integration of the elements; and the parentheses are used for the separation of "inner worlds". The Universum (including individuals who reflect it) is represented with a special polynomial. For example, at the time t_0 the Universum was:

(0) $\Omega_0 = T$,

that is, the "reality" from an external observer's point of view. Then, at time t_1, individual X "becomes aware" of the Universum:

(1) $\Omega_1 = \Omega_0 + \Omega_0 x = T + Tx$,

where Tx is the reality T from X's point of view.
At time t_2, individual Y performs an act of awareness:

(2) $\Omega_2 = \Omega_1 + \Omega_1 y = T + Tx + (T+Tx)y$,

where (T+Tx)y means that Y has an image of reality (T) and image of reality from X's point of view (Tx).
And, at time t_3, individual Z performs his act of awareness:

(3) $\Omega_3 = \Omega_2 + \Omega_2 z = T + Tx + (T+Tx)y + (T+Tx + (T+Tx)y)z$,

where (T+Tx + (T+Tx)y)z means that Z has an image of T + Tx and also of T + Tx from Y's point of view.

Now we can pose a question about the formal rules for the transformations of the Universum from one state to another in the above example. These transformations can be described as procedures of multiplying polynomials with Boolean coefficients:

$\Omega_1 = T(1+x)$

$\Omega_2 = T(1+x)(1+y)$

$\Omega_3 = T(1+x)(1+y)(1+z)$

Thus, we have polynomials of two types: those describing the states of reflexive systems and those describing the operation of awareness. (A detailed description of these problems is given in my book Structure of Awareness, 1977, which is the translation of a revised version of Conflicting Structures.)

Let us compare the expressions that depict elementary acts of self-representation in laws of form and reflexive analysis.

Laws of Form	Reflexive Analysis
$\overline{\rceil}$ =	Txx

In laws of form, this elementary act is connected to a procedure of calculation. In reflexive analysis, there is no such procedure. As a consequence, each configuration in laws of forms has its own value, but a polynomial representing the reflexive system does not. On the other hand, the syntactical structure of a formula in reflexive analysis has its own

<u>psychological interpretation</u>: Txx is "Tx from X's point of view", but ⌐|
has no biological interpretation.

Therefore, in the framework of Western cybernetics, the dominant idea
in studying systems with self-representation became the idea of
<u>computation</u>, and in the Soviet Union, the dominant idea became that of
<u>structure</u>.

4. CYBERNETIC CUBE

To represent cybernetics as an integral area of research, I will
construct a "space" of cybernetics using a cube, three edges of which are
three fundamental "ideas-coordinates": structure(X), computation(Y), and
reflexion(Z).

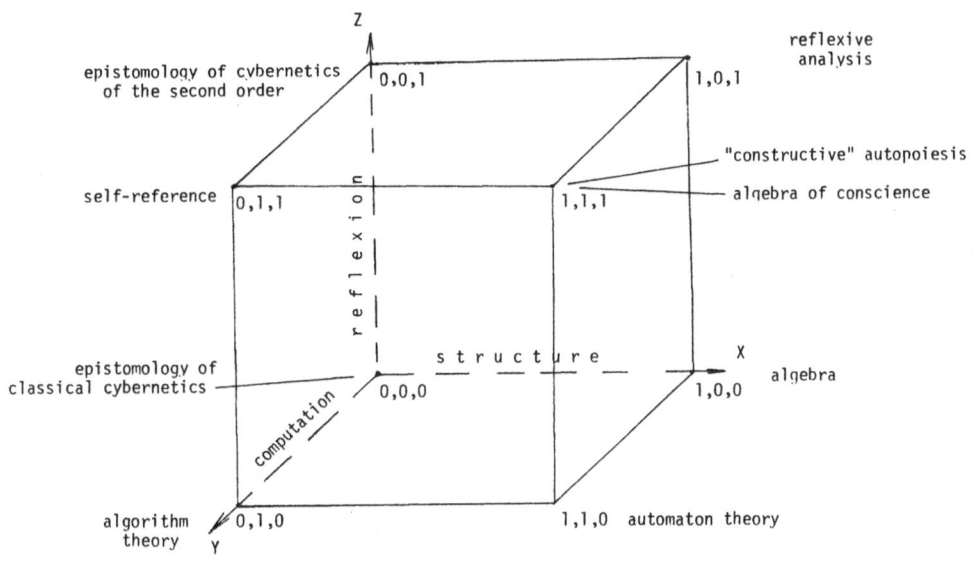

Fig. 2.

1. Traditional cybernetics corresponds to face XY. It shows the
joining of ideas of structure and computation in different ways: general
epistemological problems (0,0,0) algebra (1,0,0), algorithm theory (0,1,0)
and automaton theory (1,1,0). In this area no significant differences
between Soviet and American cybernetics exist.

2. The problems related to self-reference lie on face YZ. The concept of structure does not have an "ontological reference" to this area
(it is not used for the representation of anything which differs from the
process of computation). This area is well developed in American
cybernetics and not at all developed in Soviet cybernetics.

3. Reflexive analysis corresponds to face XZ. There, the concept of
computation does not have an ontological reference distinct from the
procedure of structures transformation. This area is well developed in
Soviet cybernetics and undeveloped in American cybernetics.

4. Second order cybernetics lies on the upper face. But American and Soviet branches are developed on different edges of the cube and they complement each other.

5. Point (1,1,1) corresponds to the synthesis of all three fundamental concepts. We can find here automata with semantics which have biological or psychological interpretations.

Thus, the appearance of second order cybernetics is the appearance of a new dimension - reflexion. However, this dimension was developed differently in the Soviet Union and the West. In the Soviet Union, the idea of reflexion was combined with the idea of structure; as a result, reflexive analysis appeared. In the West, the idea of reflexion was combined with the idea of computation; as a result, calculations with self-reference appeared.

5. SYNTHESIS

The cybernetic cube allows us to predict the future development of cybernetics: the synthesis of all the three ideas - structure, computation, and reflexion. I have made a first step in this direction by developing an "algebra of conscience". Its main idea can be seen in the following figure:

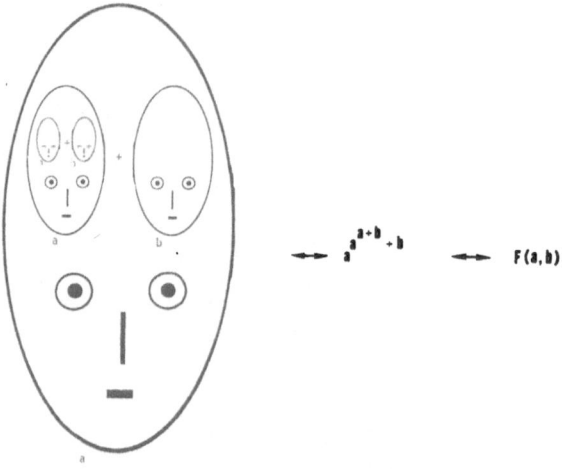

$$\leftrightarrow \quad a^{a^{a+b}+b} \quad \leftarrow \quad F(a,b)$$

Fig. 3.

The outer oval a is an individual who has images of himself (inner a), of his partner (inner b), and of their relationship (symbol +). The images also can have images (the smaller ovals), and so on. All together they constitute a reflexive structure. This structure is isomorphic to an exponential formula

$$a^{a^{a+b}+b},$$

which on the other hand represents a function. Thus, we have a graphic object (a formula) that combines a structure which can be interpreted and a procedure of computation. This method allows us to describe both the structure of an individual's cognition and his behavior at the same time. (A detailed description of this method is given in Lefebvre, 1982.)

Another step toward the synthesis of the three main ideas in cybernetics has been made by Varela (1979), Maturana and Varela (1980), and Zeleny (1980) in their development of the theory of autopoiesis, especially the part related to the modeling of biological systems.

The branches of second order cybernetics developed in the Soviet Union and the West are, as we mentioned, complementary. Consequently, the synthesis of the three ideas: structure, computation, and reflexion will also constitute the synthesis of Soviet and Western cybernetics.

REFERENCES

Baranov, P.V. and A.F.Trudoliubov, 1969a, Ob odnoi igre cheloveka s avtomatom, provodiashchim refleksivnoe upravlenie (On a game between a human subject and an automaton exercising reflexive control), Problemy Evristiki, Vysshaya Shkola, Moscow.

Baranov, P.V. and A.F.Trudoliubov, 1969b, O vozmozhnosti sozdania skhemy refleksivnogo upravlenia, nezavisimoy ot suzheta eksperimentalno-igrovoy situatsii (On the possibility of constructing a scheme of reflexive control independent from game-experimental situation), Problemy Evristiki, Vysshaya Shkola, Moscow.

Baranov, P.V., 1976, Refleksivnoe upravlenie i refleksivnaya structura resheny v igrakh dvukh lits so strogim sopernichestvom (Reflexive control and reflexive structure of decisions in two person games with strong rivalry). Problemy Priniatia Resheny, Moscow.

Baranov, P.V. and A.F.Trudoliubov, 1977, Refleksivnye protsessy v igrakh na setiakh zavisimostey (Reflexive processes in games on graphs). Veroyatnostnoe Prognozirovanie v Deyatelnosti Cheloveka, Moscow.

Druzhinin, V.V. and D.S.Kontorov, 1972, "Idea, Algorithm, Reshenie" (Idea, Algorithm, Decision), Military Press, Moscow.

Druzhinin, V.V. and D.S.Kontorov, 1976, "Voprosy Voennoy Sistemotekhniki" (Problems of Military System Design), Military Press, Moscow.

Druzhinin, V.V. and D.S.Kontorov, 1982, "Konfliktnaya Radiolokatsia" (Conflicting Radar Detection), Radio i Sviaz, Moscow.

Lefebvre, V.A., 1962, O sposobakh predstavlenia ob'ektov kak sistem (On methods of representing objects as systems), Logika Nauchnogo Issledovania (Theses of conference papers), Kiev University Press.

Lefebvre, V.A., 1965, O samo-organizuyushchikhsia i samo-refleksivnykh sistemakh i ikh issledovanii (On self-organizing and self-reflexive systems), Problemy Issledovania Sistem i Structur, Academy of Sciences Press, Moscow.

Lefebvre, V.A., 1967, "Konfliktuyushchie Struktury" (Conflicting Structures), Moscow, Vysshaya Shkola; second edition: 1973, Sovetskoe Radio, Moscow.

Lefebvre, V.A. and G.L.Smoljan, 1969a, Algebraische Darstellung menschlicher Konfliktsituationen, Ideen des exakten Wissens, No.1.

Lefebvre, V.A., 1969b, Janus-Kosmologie, Ideen des exakten Wissens, No.6.

Lefebvre, V.A., 1970, Das System im System, Ideen des exakten Wissens, No.10.

Lefebvre, V.A., 1972, A Formal Method of Investigating Reflective Processes, General Systems, Vol.XVII.

Lefebvre, V.A., 1973, Auf dem Wege zur psychographischen Mathematik, Ideen des exakten Wissens, No.6.

Lefebvre, V.A., 1975, Iconic Calculus: Symbols with Feeling in Mathematical Structures, General Systems, Vol.XX.

Lefebvre, V.A., 1977, "The Structure of Awareness: Toward a Symbolic Language of Human Reflexion", SAGE Publications, Bevery Hills.

Lefebvre, V.A., 1982, "Algebra of Conscience: A Comparative Analysis of Western and Soviet Ethical Systems", D. Reidel Publishing Company, Dordrecht, Holland.

Lepsky, V.E., 1969, Issledovanie refleksivnykh protsessov v eksperimente na matrichnoy igre s nulevoy summoy (Investigation of reflexive processes in an experiment on a zero-sum matrix game), Problemy Evristiki, Vysshaya Shkola, Moscow.

Maturana, H.R. and F.Varela, 1980, "Autopoiesis and Cognition: The Realization of the Living", D. Reidel, Holland.

Schreider, Yu.A., 1973, K postroeniu yazyka opisania sistem (On the construction of the language of systems description), Systemnye Issledovania, Nauka, Moscow.

Schreider, Yu.A., 1975, Slozhnye sistemy i kosmologicheskie printsipy (Complex systems and cosmological principles), Sistemnye Issledovania, Nauka, Moscow.

Schreider, Yu.A., 1983, Osobennosti opisania slozhnykh sistem (Peculiarities of complex systems description), Sistemnye Issledovania, Moscow.

Spencer Brown, G., 1969, "Laws of Form", George Allen and Unwin, London.

Toom, A.L., 1973, Sposoby priniatia resheny v odnom klasse igr (Analysis of decision making in a special class of games), Izvestia AN SSSR. Tekhnicheskaya Kibernetika, No.3.

Toom, A.L. and A.F.Trudoliubov, 1974, Reflexive Wechselbeziehungen im Kollektiv, Ideen des exakten Wissens, No.3.

Toom, A.L., 1976, Nesimmetrichnaya kommunikatsia, fokalizatsia i upravlenie v igrakh (Non-symmetrical communication, focalization, and control in games), Semiotika i Informatika, Moscow, No.7.

Toom, A.L., 1978, O roli znakovoi situatsii v igrakh (On the role of the semiotic situation in games), Semiotika i Informatika, Moscow, No.10.

Toom, A.L. 1981, Na puti k refleksivnomu analizu khudozhestvennoy prozy (Toward reflexive analysis of fictional prose), Semiotika i Informatika, Moscow, No.17.

Trudoliubov, A.F., 1972, "Reshenia na setiakh zavisimostei i refleksivnye mnogochleny" (Decisions on graphs and reflexive polynomials), <u>IV simposium po Kibernetike</u>, Tbilisi.

Varela, F., 1975, A Calculus for Self-Reference, <u>Int.J.Gen.Systems</u>, Vol.2.

Varela, F., 1979, "Principles of Biological Autonomy", North Holland, New York.

von Foerster, H., 1981, "Observing Systems", Intersystems Publications, USA.

Zeleny, M., ed., 1980, "Autopoiesis, Dissipative Structures, and Spontaneous Social Orders", Western Press, Boulder, CO.

METHODS FOR MAKING SOCIAL ORGANIZATIONS ADAPTIVE

Stuart A. Umpleby

In Robert Trappl's opening remarks on the first day of this conference he raised the issue of the usefulness of the theories that we debate
with each other at these conferences every two years. Stafford Beer in
his address made a similar point when he suggested that we confront the
way things are. I follow their lead by suggesting that we really know
quite a lot about how to solve social problems and how to make social organizations more effective. But for some reason we are not using the
knowledge we have. Why we do not make better use of our current knowledge
is the issue that I would like to explore. My method of exploring will be
to investigate the history of ideas in the field of cybernetics and
general systems theory.

ASHBY'S THEORY OF ADAPTATION

Several people have made reference during this conference to the work
of Ross Ashby. I, too, have found Ashby's work to provide an
indispensable foundation for further work in cybernetics. Hence, I shall
begin with a reference to Ashby's theory of adaptation (Ashby, 1960).
Recall that he proposed that any system with two nested feedback loops
would be capable of displaying adaptive behavior. Let me give an example.
When you drive an automobile, you have to make a large number of small
corrections to keep the car on the road, to avoid hitting pedestrians, to
stop at traffic lights, etc. However, if you are driving a regular route
between your home and office, you can do all of these things almost
without thinking. In fact you may have had the experience of driving
along a familiar route and neglecting to turn off where you intended,
because you were thinking about something else. In a sense you were
driving on "automatic pilot" down a familiar road.

But periodically you may have to find a new route. Perhaps you have
changed either your home or office address, or perhaps the city is doing
some repair work on the road. If so, you will have to find a new pattern
of routine behavior. In Ashby's scheme the first feedback loop (keeping
the car on the accustomed route) operates frequently and requires making
only minor corrections. The second feedback loop (finding a new route)
operates less frequently but restructures the routine behavior in a major
way. The overall behavior can be said to be adaptive. The driver not
only regularly arrives at a predetermined goal but also is able to learn
new patterns of behavior to achieve the goal in a changing environment.

Large social organizations such as corporations and government
agencies also display adaptive behavior. They take many small corrective
actions each day but only occasionally make a major change, such as

introducing a new product or buying or selling a subsidiary firm. The question that I would like to ask is: Have we given sufficient attention to BOTH feedback loops? It seems to me that most of the work that has been done in cybernetics and systems theory has tended to focus on small corrections in organizations. I think that we have done a very good job of advancing the technology for managing day-to-day operations. But our efforts in this area seem to be yielding diminishing returns. The gains have not been as great in recent years as they were in the years immediately after World War II. The kinds of difficulties that American managers are facing today tend to be less technological than cultural. That is, there is at least as much attention being focused now on how to get people to use existing technology as there is on developing additional technology.

THE INTERACTION BETWEEN IDEAS AND SOCIETY

When one attempts to make a major change in an organization, such as the adoption of a new technology, one encounters a great deal of resistance. People have established customs and patterns of behavior that are familiar and are not given up easily. The kinds of approaches that are usually found within the field of operations research or the mathematical decision sciences do not address the question of how one achieves widespread support for a major transformation of a social organization. Although there has been some very nice work on organizational structure and the ideal design of a viable system, a question that remains is how one achieves widespread support for the change from the existing structure to the preferred structure.

One of the truths we tend to forget is, "What we think determines how we act, and we can change the way we think." Let me give several examples to illustrate the point. The examples come from science, everyday life, international relations, and management consulting. In the case of science there is a fundamental difference between the natural sciences and the social sciences. When a major change in theory occurs in the physical sciences, we assume that only the theory changes. The behavior of objects does not change. When physicists changed their principal theory from classical mechanics to modern quantum mechanics, the behavior of atoms did not change. But in social science, the situation is quite different. In fact, one of the reasons we formulate theories of social systems is that we want our social systems to operate differently.

In the case of everyday life we also understand the interaction between ideas and circumstances. For example it is possible to have a small income and yet be content as long as one's wants lie within one's resources. However, it is also possible to have a large income and not be content, that is, if one's reach exceeds one's grasp.

In the case of international diplomacy there are a number of successful applications of theories which assume an interaction between ideas and behavior. One example was the effort by President John F. Kennedy during the period following the Cuban missile crisis to obtain a Nuclear Test Ban Treaty. Between October 1962 and the summer of 1963, a period of about nine months, Kennedy, obviously with the cooperation of Khrushchev, was able to achieve a major reversal in the climate of Soviet-American relations by applying a theory of graduated and reciprocated initiatives in tension reduction. (Etzioni, 1967)

In the case of management consulting, we know a great deal about how to improve our work environments. There are some organizations that are dynamic and innovative. There are other organizations where people

134

complain a lot. When people are unhappy, they are not as productive. Consequently methods have been developed for transforming a less productive organization into a more productive organization.

The point is that we do have the means to transform our social systems. Of course individuals and societies must change at their own pace. Nevertheless, methods do exist for making major changes in a fairly short period of time. Perhaps I should give one or two examples of the methods that I have in mind. I shall then describe the assumptions that these methods are based upon. I think you will see that a theoretical foundation for these methods would entail a major departure from the theories and epistemologies that underlie most of the papers we present at conferences of this sort.

DESIGNING DISCUSSIONS TO RESTRUCTURE ORGANIZATIONS

If we accept the idea that the way to change behavior is to change the way that people think and the idea that organizations are composed of many independent human beings, then the task is to lead the people in an organization through a discussion so that they jointly arrive at a new, more productive pattern of behavior. Numerous methods have been divised to lead such a group discussion. I shall briefly describe just two of them.

The first method for conducting a group discussion or a planning meeting within an organization is called LENS, which stands for Leadership Effectiveness and New Strategies. (Umpleby, 1983) The method has been developed by the Institute of Cultural Affairs, a community development organization based in Chicago. The method involves leading a structured discussion. An activity of this kind is somewhat analogous to conducting an orchestra. First one writes the score, and then one leads the playing of the score. In the case of LENS the group proceeds through a discussion of their ideas about five things.

(1) Their vision of the future. What they want for themselves, their families, and their community or organization.

(2) The contradictions or obstacles to achieving the vision. If people agree upon a common vision, why do they not have it? What are the factors that are impeding progress?

(3) Programs to remove the obstacles to achieving the vision.

(4) Tactics for implementing the programs.

(5) Who will do what, when, where, and how in order to carry out the tactics.

The LENS method is usually performed during a five day conference. However, shorter versions can be done in less than one day. Regardless of the length of time devoted to the meeting, some time is spent together as a single group and some time is spent in smaller groups, so that everyone has an opportunity to speak. Expanding participation both generates additional ideas and builds commitment to the plans that are eventually formulated.

The method developed by Russell Ackoff and his colleagues for dealing with corporate clients is similar. (Ackoff, 1981) Their method also involves a series of discussions. However, rather than starting with a vision of the future and then looking at the obstacles to achieving it,

Ackoff assumes that the vision in most organizations is to continue doing what they are doing. He feels that in order to convince corporate managers that change is needed, it is necessary to show them that their current policies will eventually lead to collapse and ruin. Hence in Ackoff's method the first step is to project current behavior in order to identify the mode of future collapse. Once the future failure hidden within current practices is identified, it is then possible to create an idealized design that will avoid that future problem or set of problems. Subsequent steps involve ends planning and means planning. Ackoff uses technical analysis more than the Institute of Cultural Affairs which focuses on training and motivation. Using our musical analogy again, the difference can be thought of as the difference between two composers. Both are designing a pattern of behavior for a group of people to carry out. In each case the result of the planning activity is a new pattern of organizational behavior.

There is an emotional component to this kind of activity regardless of who the composer is. Hence, a successful composition must also deal with this aspect of the group's interactions. To do this, one must be aware of, and at least in part redesign, the myths that exist within the organization. For example, the Institute of Cultural Affairs writes songs about the organization that summon up pride in the past, hope for the future, good will toward fellow workers, and a determination to overcome obstacles. The Institute also uses symbols that remind people of the feelings that they have when they are working successfully together.

HOW NEWER METHODS DIFFER FROM OLDER METHODS

Now let us look at how these methods, which have been successfully used in communities, corporations, and government agencies, are different from the usual mathematical applications of cybernetics and systems theory.

The methods I have described have not had their origins in the purely academic world. They are the result of practice. They have been developed by people who were trying to get things done in organizations. Here is a list of some of the characteristics or underlying assumptions of the LENS method. The LENS method differs more sharply from classical decision science than does the work of Ackoff and his colleagues.

(1) The problems are virtually never clearly defined. There is a set of interrelated problems.

(2) The focus is on action at least as much as analysis, that is, in moving in a positive direction, even if one does not have a complete analysis to begin with. The reason lies in the need to restore or maintain hope that positive change can take place.

(3) There is more attention paid to getting people to work cooperatively together than to devising an optimal solution. Hence the focus is on psychological constraints more than physical constraints. Rather than optimizing within physical constraints, the task is to remove psychological constraints.

(4) Regarding data it is assumed that the people taking part in the meetings have sufficient data in order to act. Usually one tries to get people outside the organization to take part in order to bring fresh perspectives and additional skills.

136

(5) The procedure for implementing the plan is part of the plan that is developed. One does not solve a problem and then hand the solution to a decision-maker. Rather, the people who will have to implement the plan are the people who formulate it.

(6) The emphasis is not on using an analytical technique but rather on improving communication among the people involved. The central concerns are motivation, organization, and cooperation and how to achieve them.

Now the question is, since these methods are rather widely known, why do we not pay more attention to this kind of work? I think that an important reason lies in our epistemology. In order to understand theoretically what is happening when these methods operate, we have needed new concepts that we did not have before. Some of the necessary concepts are presented in the papers in this volume. First of all, we need to think of individuals as autonomous elements. We need to think of a social system as a collection of autonomous elements where each one of those elements creates its own vision of the world. People have their own views of their personal goals, the goals of the organization, and how they want to interact with the organization. Those ideas and aspirations need to be included in the plans that are developed. Also, we need to think of knowledge not as something in the mind of an expert, but rather as some-thing that is shared by a group of people.

Contrary to this notion of a social system as a collection of autonomous individuals constructing their own realities, decision scientists have tended to think of social systems as a set of interacting variables. We have asked questions such as, "If variable A increases, what will happen to variable B?" Once a model was constructed, our concept of implementation was to make a recommendation to a decision-maker who would then issue the necessary commands. This conception is more compatible with Ashby's first feedback loop (small day-to-day corrections) than with Ashby's second feedback loop (a major change in the pattern of behavior). The new methods have been designed to deal with major structural change.

A more appropriate question from the perspective of the new methods and the new epistemology is to ask, "How can we design a discussion so that people come to agreement on topics that will improve the functioning of the organization?" Our older mathematical methods constitute a powerful technology, but they are a long way in their basic assumptions from the idea of a social system as a set of interacting, autonomous human beings. The lack of an epistemology compatible with the newer methods is probably an important part of the explanation for why the newer methods are not more widely used.

When we use the methods that I have described, we shift from being a technical analyst to being a person who composes and then conducts a discussion among a group of people. There seem to be two routes whereby one can come to an understanding of the ideas presented in this symposium. One route is to do client-based work. The other route is through epistemology. The two areas of inquiry are mutually supportive. With the work that has been done in the area of second order cybernetics, it is now possible to have a more thorough theoretical understanding of the kinds of group process methods that have been shown to be effective.

REFERENCES

Ackoff, R., 1981, "Creating the Corporate Future", J. Wiley, N.Y., London, and Sidney.

Ashby, W. R., 1960, "Design for a Brain", Second Edition, Chapman and Hall, London.

Etzioni, A., 1967, The Kennedy Experiment, The Western Political Quarterly, Vol. XX, No. 2, Part 1.

Umpleby, S.A., 1983, A Group Process Approach to Organizational Change, in: "Adequate Modeling of Systems", Horst Wedde, ed., Springer Verlag, Heidelberg and Berlin.

DISCUSSION: GUIDING QUESTIONS AND CONCEPTUAL STRUCTURES IN

CYBERNETICS AND GENERAL SYSTEMS THEORY: COMPARATIVE STUDIES

Moderator: Stuart A. Umpleby

Umpleby: We have quite a bit of time for discussion.   (Addressing Ernst
    von Glasersfeld:)   Ernst, did you have something you wanted to say?

Von Glasersfeld: Well, I just asked whether we could get back to the con-
    cept of information because that was brought up and that is a whole
    bag of interesting things ....

Umpleby: Seeing no comments on information, are there other questions?

Beer: Regarding the distinction between psychological and logical self-
    reference, I always wonder about that difference. You (turning to
    Lefebvre) talked about it. I have never been in Russia; I bumped
    into this problem in a series of meetings we organized on self-refer-
    ence. It is very strange that there is an incompatibility there.
    Some people will approach self-reference in a logical way, and some
    other people will approach it in a psychological way. I find it very
    difficult ever to get that discussion to merge. So I would be inter-
    ested how you think the two relate to each other?

Lefebvre: I shall give you an example. Consider the social or cultural
    picture of a saint. The question is: does a saint know that he is a
    saint? Of course, if a saint thinks that he is a saint, from a cul-
    tural point of view he is not a saint. So, whether a person can be
    correctly thought to be a saint depends on who the observer is. In
    order to describe itself, an organism must contain an image of itself
    (logical self-reference). But a saint, in order to be a saint, will
    not think that he is a saint (psychological self-reference).

Beer: Are you aiming at the difference between the biological and the so-
    cial sciences?

Lefebvre: Biological and cultural-cognitive sciences.

Umpleby: I have a question for Ernst von Glasersfeld. I would like for
    you to say more about how you can get values from the assumptions you
    make about your epistemology. I did not thoroughly understand what
    your point was, but it sounded fascinating. Could you expand on
    that?

Von Glasersfeld: I will try. The assumption is that a constructivist
    epistemology is relevant to this question. That is, the experiential
    reality that each cognitive organism has of himself or herself is

constructed on different levels, different 'degrees of reality', if you like. I briefly outlined 'repetition', 'multi-model confirmation' and as the last level what we called 'corroboration by someone else'. Now, since we all tend, I think, to want that last level, we want a reality that is more and more reliable. Therefore, the more confirmation we can get, the better. So we tend to look for this last level. The last level requires the construction of others. Therefore, the others become essential in our construction of reality, and this essentialness proposes an ethics. Because we need the others, we are dependent on the others.

Umpleby: ... for confirmation ....

Von Glasersfeld: ... for the construction of our own reality and making it more stable and more reliable. When Kant in his Categorical Imperative says : 'Act always in such a way that the maxim of your action can be taken as a maxim by others', I would turn that around - to say that I turn it around is perhaps not quite right, but I look at the other side of that. That is, in order to decide how we can act, we have to devise a way of acting that we then can attribute to others so that we can predict them. This puts constraints on our own actions and our own acting.

From the floor: If there are different or 'other' realities as Prof. Beer supposed, I would say that there might be quite different systems and the question will be whether they could co-operate. Reality, as I understand you, can only be defined from the process of perception and model-making and so on. If there is another person, another reality, how can the other person interact with the first one?

Von Glasersfeld: I would answer by saying that at the beginning of my talk I tried to stress the fact that I am separating epistemology from ontology. What I am talking about is the cognitive level and whatever happens on the cognitive level. Now, when you say that there are other realities, you are talking ontologically. As far as I am concerned my experiential reality is constructed out of my experience but nothing, nothing else. Thus, and I should stress that every time I open my mouth, I am not talking about what exists in the world, I am not talking about being, I am not trying to give a description of anything. I am trying to construct a model that allows me to think about my own cognitive processes and their results.

From the floor: Is it necessary that these results are unique or without any contradictions?

Von Glasersfeld: How do you mean 'unique'?

From the floor: By 'unique' I mean that your model is conclusive.

Von Glasersfeld: I don't like the word unique. I would say that it is necessary, or it is an assumption of mine that it is necessary, that the model becomes consistent. So, if you want to talk in the traditional, epistemological terms, I accept the criterion of consistency, but not the criterion of correspondence.

From the floor: I just want to add to this point. In modern ethnology we have the problem of describing what other people in other cultural domains are doing and how they understand the world. Until now we always tried to implement our own world-view into theirs and we could not understand what they were doing; e.g. we said witchcraft is psycho-pathological. But in their context it is not. It is just as rational as our building cars. But it seems that both views are not compatible, cannot be compared with each other. What would your model say about this? I think it would say that those persons cannot interact with each other.

Von Glasersfeld: Well, they can interact. But let us put the discussion in very practical terms. Suppose I go to some island where I have never been before. There is a tribe. I try to understand what they are doing. I cannot do anything but interpret what I experience of them - how I see them act; what I see them do - in my terms. Now, when I say 'my terms' that does not necessarily mean that I explain it analogously to what I am doing, but it has to be something that I can see myself possibly doing. It has to be within the potential that I ascribe to myself. I can never know what they are thinking about their own actions. That, I think, is what the origins of hermeneutics were. The early work was extraordinarily brilliant and interesting but, when hermeneuticians try to bring some objectivity into that, they destroy everything that hermeneutics was originally trying to do.

From the floor: About ethics, I think that your ideas do not explain anything but simple egoism. They result in simple egoism. I do not understand how you could ever come to the conclusion that if I as a person construct my world from inside with reference to other persons, and I need the existence of other persons only as far as they contribute to my self-creation or self-organizing, then how can I ever think that they have rights against me, that I could accept that I have to suffer laws or do something for them, if it means a subjective loss to me. I do not see how you can jump from this kind of egoistic self-centered view to a generalized ethical perspective or duty. I just can't see it.

Von Glasersfeld: Obviously that would require a very long answer. Even my paper is probably too short to answer all your questions, but let me say this: To me that is on the same level as my construction of the physical world. I do believe that I construct my physical world, but I have to construct it within certain constraints. I cannot now turn around and walk through here (through this wall) into Vienna. That is reality, but what is reality? What do I know of it? What I know of it is exclusively that my way of acting - my way of thinking of it - has certain constraints. It does not allow me to do that. It does not tell me what the wall is. It does not tell me what Vienna is in an ontological sense. I know it through my experiences, whether they are sensory or whether they are conceptual. Thinking experiences are breakdowns of certain things that I am doing. Warren McCulloch a very long time ago said that the peak of knowledge was the breaking down of a hypothesis. I buy that hook, line, and sinker. That is the basic notion that ties my kind of knowledge to ontological reality. Now regarding your second point - that what I am doing is nothing but pragmatism. First of all, I do not pretend to be doing anything new, but the pragmatists by and large still showed somewhere a hankering, a nostalgia for iconic knowledge. If you read Charles Pierce, you will find that he has absolutely wonderful things in his works. But he also wants to be a realist. He wants to believe that if you do that long enough you come to something that is very like a picture of the world. That is what I am trying to eliminate, because I think that it is unnecessary. I think that it is confounding, and it leads people up the garden path.

From the floor: What about the wall?

Von Glasersfeld: The wall is my construct - of course it is. And it is a very useful one, because I don't bump my nose once I have constructed the wall.

From the floor: Why don't you construct it in such a way that you can walk through it?

Von Glasersfeld: Because there is an ontology, but I cannot express or describe that ontology in other ways except the failure of my actions. Let me add one thing that may help. Have you ever seen a blind man walk with his stick? This is an old metaphor. What does that blind man see of his surroundings? Only the impacts on his stick. He can draw a map out of that, and after a while he learns to move in that landscape quite efficiently. But that landscape for him consists of nothing but points of resistance.

Sadovsky: It seems to me that if I try to find the historical sources of the approaches which Professor von Glasersfeld explained to us here, I find them not in pragmatism. I agree with you, but it seems to me that your conception is about knowledge as a special kind of social activity. And in this respect I would like to ask you one question. What is your opinion about the way to solve the problem of inter-subjectivity regarding the pictures of reality which your people will construct during this kind of activity?

Von Glasersfeld: Well, the problem of intersubjectivity is a problem of the viability of my models of other people. What do we mean when we say we understand someone else? What do we mean when we agree? In my view what we are really saying is that I have now succeeded in constructing a model which, in the situations in which we have been, has been compatible with yours. We do not know whether the next moment, the next day or ten years from now your model and my model will turn out to be incompatible. Anyone who has been married for some time knows that extremely well. There is a moment, perhaps six months into the marriage, when you think: 'Now we've got it, we are going to agree about everything, we really think the same way.' Well, sure enough, after some time you realize that 'No, she is not like that' (I am talking from the male point of view). There is something else about her that I never built into my model at all. Maybe it makes me reconstruct the model I have. Maybe it makes me add another balcony or chimney or something. So intersubjective validity is always a question of compatibility, not of matching. It is 'fit' as opposed to 'match'. And I think a lot of disasters in this world have happened because people have demanded a match and have come to expect it. Therefore, when the tomorrow comes and there is a discrepancy, they say 'you've changed,' or 'the world has changed'.

Sadovsky: Yes, that is true ....

Von Glasersfeld: Sorry, let me interrupt you. It is not true. I would never say that. It is a way of thinking about the world. It is not true. God help me!

Sadovsky: I understand you. It seems to me that the final criterion is the corroboration that you mentioned many times, and I wonder if it is possible. We know very well that - as Popper said many times - if you try to find corroboration, you will find it everywhere. That is why the problem of intersubjectivity is even deeper.

Von Glasersfeld: Popper has a totally different view of corroboration. Popper, whom I respect very much, knows nothing about language. Popper still has the illusion that language is something that sort of transfers the meaning from me to someone else and therefore corroboration is somehow: 'Aha, we've got the same meaning', and the meaning is there. That, from my point of view, is nonsense. Because language operates on a principle of fit. I understand what you say if I can construct something that fits whatever constraints there are in the rest of what you say and in the situation and in the context.

Sadovsky: But the epistemology of Popper is very close to the idea of fitting. Therefore, I think that the use of Popper is rather close to finding a way to construct reality.

Von Glasersfeld: No, I have said that I have great admiration for Popper. Popper has given the best history and definition of instrumentalism, and it is a very valid one. At the end of it, when he wants to argue against it in 'Conjectures and Refutations', what does he say: 'Well, I don't believe scientists should think that way. It is bad for science.' That is not an argument.

Varela: Let me add one more aspect to this debate, which is quite fascinating. I think I disagree with you, Ernst, about fitting. Fitting is a terribly unfit word because it actually preserves exactly the notions that we want to replace. I think the clearest place in which one can see this is in evolutionary theory. If one continues to think that evolutionary mechanisms have anything to do with the fitness of one organism to certain structures of the environment, there is little we have understood about evolution. That is an input-output theory of evolution: the selective pressures come into the genetic system, and out comes a transformation of the genetic system in order for it to fit. What I want to say is that what happens in evolution is an encounter between an extremely intraconnected system which is the genetic and the ontogenetic system of an organism, and certain minimum constraints that need to be satified, e.g. it cannot live beyond 45 C or below 25 C or something like that. But once constraints are satisfied, anything goes. That is not fitting. The word I would like to suggest is 'drifting' rather than fitting because there is no match particularly. It is as if you define a broad boundary, and within it you let this autonomous system go by itself. This is not purely a problem in intersubjective epistemology, but it can be brought down to a very simple level of, let's say, cellular interaction or organismic interaction. If you try to account for evolutionary theory in terms of a particular fit, as in the classical notion of selective fitness between a genetic endowment and the world, it just does not work. Much the most interesting part of evolution, is how the internal coherences or the entire genetic and epigenetic system actually lay down a path which then becomes a species. The form of the snail or the nose of the orang-utan had nothing to do with a particular fitting to anything at all. They have to do with the fact that the genetic system is extremely tightly interconnected, so you pull it one little bit here and the whole thing changes. That the whole thing changes means that one sees the organism actually carving out a new, different world for itself. So, fitting in that context is to me the same as not bracketing objectivity properly.

Von Glasersfeld: What you say shows that we interpret the word 'fitting' differently. For me, the crucial thing in 'fit' is that when I say one thing fits with another, they have no points in common, whereas when I say they match, they match to the extent to which they have points in commmon. Now, I do not mean 'fit' in the sense that an outside observer might call the 'best fit'.

Varela: There is a history of words.

Von Glasersfeld: I know, I know that Darwin himself talked of the 'fit-test', and that superlative was very bad for him and for everyone who followed.

Varela: Yes, so when you talk about 'fitness' you immediately invoke as it was brought up here ....

Von Glasersfeld: No, Francisco, I never talked of 'fitness'. I talked of 'fitting' as a verb.

Varela: That is my case.

Von Glasersfeld: I wish I had a better verb, but if you have ever sorted gravel with a sieve, anything that goes through the sieve fits. No matter what shape, no matter what size. I think that fits your no-tion of evolution, where you have broad constraints and within the constraints you can do whatever you want to do or whatever the system develops on its own.

From the floor: I think that the essential question, from a realistic epistemology, is to understand that information comes from outside the system, it comes from the environment of the system, because a closed system cannot generate information. In fact this is a princi-ple in thermodynamics. About the other issue, I think that knowledge is not always a social activity, but that scientific knowledge <u>is</u> a social activity.

Varela: I probably express myself very poorly. I never talked about a closed system, never. To me that is an artificial construct that has little to do with science. To talk about a completely closed system is a 'Gedankenexperiment' at best. So, I never talked about a closed system. We entirely agree. Information comes from the outside. The question is: is that idea useful? From what we have said here, at least I think from what Ernst and I have said here, this is not a useful notion most of the time. Certainly it is not useful to under-stand cognition on the molecular or on the neuronal level. Certainly it is not useful to understand evolution. Some others would claim that it is not useful to understand societies as well. That is the point. Precisely because information comes from the outside, that is a type of description of the system which does not account for what we actually see it doing, namely specifying a sense for its world which does not come from the outside, but comes from the fact that there is an encounter between its structure and the constraints it finds. Within that situation it produces a path which is unique to that particular system. It has nothing to do with the second law of thermodynamics which talks about non-interacting systems. We are not talking about that. We are talking, though, about different ways of understanding how interaction takes place. What I said in my charac-terization of autonomous systems is that we should change the notion of relationship. In both cases there is a relationship with the en-vironment. In the first case it is an instructive one. In the second case it is a laying down of sense.

Question: ... organization in the system accounts for an interchange ....

Varela: Not at all. I think that that is a myth that we have inherited
     from physics. The organization of the system has to do with the fact
     that you have internally interconnected elements. The surprise is -
     I take it as a fascinating surprise - that the more we study systems
     which are tightly interconnected, this tightness of interconnections
     reveals the emergence of these eigen-behaviors as a universal quali-
     ty. And there are simple examples that can be studied mathematically
     in some cases, or in simulations, or in biological systems. But this
     has nothing to do with entropy. This system has nothing to do with
     thermodynamic principles. Of course, as a limit condition all sys-
     tems will have to satisfy the second law. So we stop there. We drop
     that and get into the interesting business.

From the floor: I think the important concept is the concept of 'in-for-
     mation'. You formerly said that this kind of information is non-lo-
     cal and I think another important distinction is the concept of cor-
     relation in connection to information, e.g. in physics. Most physi-
     cists do not understand, for instance, the EPR paradox because they
     believe that there is a transfer of information. But there is not.
     There is only pure correlation. I think that in highly complex sys-
     tems we have to look for correlations and not for information flow or
     something like that. Shannon-type information expresses itself in
     correlations, and I think that the search for correlations in such
     systems should be enhanced. There is another moral behind it.   You
     need not look for causal changes. You get information from the sys-
     tem in quite another way.

Varela: This is exactly the point. Thank you very much.   I think you
     have said it very well, but I just want to stress that this is no ab-
     stract notion - that this can be put to work. Many times we are just
     not aware how useful it can be to take that route. In genetic sys-
     tems this is very clear. In the genetic system you have the classi-
     cal notion that we were taught in school which is that you have one
     gene having to do with one character which has to do with one aspect
     of the environment.   At the same time every gene depends on every
     other gene, so that whatever a gene does will be an expression of the
     non-locality of the situation. From that moment onwards you immedi-
     ately realize that the internal state of a genetic network is not
     anything that is, in the classical sense, in 'perfect match' or
     agreement or fitness in the Darwinian sense with the environment. It
     is simply at one state which is made compatible with minimal
     constraints. Once you have that point, you can understand why it is
     that you have the frequency of genes in a certain population and why
     it changes the way it does. If you do not drop the other notion,
     there is simply no way to understand the data. This is to me the
     most important point that I keep coming back to. There are concrete,
     specific problems that cannot be solved unless we are willing to make
     that change of point of view - bracketing the objectivity and staying
     with an autonomous point of view. It is not merely of epistemologi-
     cal interest. It is also of a very pragmatic and direct interest.

From the floor (addressing Varela): You contrast these two concepts of
     transfer function on the one side and, as you call it, operational
     closure on the other. Are these completely distinct concepts or are
     they ideal types in the sense of Max Weber - the abstract ideal forms
     - or are they maybe two ends of a continuum? For example, you indi-
     cated as transfer functions the input-output relation. I would like
     to mention that this input-output relation becomes less and less im-
     portant and the internal state becomes more and more important, and
     finally you end up with your questions. Is this right or is this
     wrong?

Varela: Well, I do not take it as a dogmatic decision that it should be this way or should be the other way. To me it is very much an open question that will have to be looked at and decided on the basis of future developments. I would say, let us wait some 15 years or so until we have fully developed the theory of second-order cybernetics or autonomous systems and then see if we find that there is some intermediate basis. Maybe it is an all or none situation. So, to me it is very much an open, empirical issue. I don't know. Or I would not like to say what it is a priori.

From the floor: You mentioned something earlier, but you came back to it now. You mentioned the difference between the Turing machine approach and the autonomous approach. I believe you said that the Turing machine is universal in the sense that it is universal in representation. You can represent anything. And of course autonomous systems can also represent anything. So I would think that there can be a formal equivalence, and the only difference between the two would be that one representation is more suitable to understand one kind of systems or one kind of processes and the other more for another kind of processes. I would like to see how you translate from one representation to the other so that if you have some problem in one representation you can see that it is a pseudo-problem from the other point of view. I think there is a kind of formal equivalence, but as to our cognitive abilities we must use the appropriate one.

Varela: I think it is very clear what you said. I do not have an a priori idea of the issue. But if you push me, as you just did, I shall offer a conjecture. My conjecture is not that they are equivalent, but that one is, at the limit, reducible to the other one. In other words, when you reduce an autonomous machine to its minimal form, it becomes a Turing or a Turing-like machine. It is more like general relativity and Newtonian mechanics. Not that one contradicts the other, but one reduces to the other in the limit. My hunch is that if we had a clear mathematical formulation of autonomous machines, then that issue could be decided on a more sound basis. At this point we only have fragments. There is a lot of work that needs to be done. Where I would do research is in showing that one is reducible to the other one at the limit. The general class is the autonomous, but when you reduce it to its minimal form, it becomes Turing-like.

From the floor: Let us say you have machine A and machine B. If you can reduce A to B, can you go the other way back - from B to A? Is it reducible both ways? Is that what you are saying?

Varela: If the two classes of systems have the same status of explicit mathematical formalism, then we could see whether we have a bijective morphism in both directions, in which case there will be a class of complementarities. The other situation is where we could have the morphism only in one direction. That is an open question.

From the floor: So, if you have the first situation, then they will be equivalent?

Varela: No, not equivalent, complementary.

From the floor: How do you define complementary?

Varela: Complementary means that you can have either this one or that one. They are transferable, but they are not necessarily reducible to each other. This is a hunch, not a proof.

From the floor: Is complementarity comparable to complementarity between autonomy and reliability?

Varela: And reliability?

From the floor: Yes, reliability. Let us say the Turing machine corresponds to reliability in the sense of predictability and computability and the autonomous machine corresponds to a surprise-behavior.

Varela: That is one dimension in which they could be seen as complementary.

Umpleby: We have been talking for two and a half hours, I suggest that other questions be directed to people privately. Thank you all.

BIOGRAPHIES OF CONTRIBUTORS

STAFFORD BEER is an international consultant in the management sciences. For twenty years he was a manager himself, and has held the positions of company director, managing director, and Chairman of the Board. He is currently a director of the British software house, Metapraxis Ltd.

In part-time academic appointments, he is visiting professor of cybernetics at Manchester University in the Business School, and adjunct professor of social sciences at Pennsylvania University in the Wharton School, where his previous position was in statistics and operations research.

He is President of the World Organization of General Systems and Cybernetics, and holds its Wiener Memorial Gold Medal.

His consultancy has covered small and large companies, national and international agencies, together with government-based contracts in some fifteen countries. He is cybernetics advisor to Ernst and Whinney in Canada.

Publications cover more than two hundred items, and nine books. He has exhibited paintings, published poetry, teaches yoga, and lists his recreations as spinning wool and staying put in his remote Welsh cottage.

Address: Prof.Stafford Beer
        Cwarel Isaf
        Pont Creuddyn
        Llanbedr Pont Steffan
        Dyfed SA48 8PG
        UK

ERNST VON GLASERSFELD was born in 1917 of Austrian parents, went to school in Italy and Switzerland, briefly studied mathematics in Zuerich and Vienna, and survived the war as a farmer in Ireland. In 1948 he joined the research group of Silvio Ceccato who subsequently founded the Center for Cybernetics in Milan. In 1963 he received a contract from the U.S. Air Force Office of Scientific Research for work in computational linguistics, and in 1966 he and his team moved to Athens, Georgia. Since 1970 he has taught cognitive psychology at the University of Georgia. His principal interests are conceptual analysis, epistemology, and the development of number concepts in children. He is currently working on a book on the constructivist theory of knowledge.

Address: Prof.Ernst von Glasersfeld
        University of Georgia
        Dept. of Psychology

180 Shadybrook Drive
Athens, GA 30605
USA

VLADIMIR A. LEFEBVRE was born in the USSR and received his diplomas from Moscow State University: first at the mechanics-mathematical department (equivalent of M.S.), then at the psychology department (equivalent of Ph.D.).

His book Conflicting Structures was published in Moscow in 1967 and 1973. An English version of this book entitled The Structure of Awareness was published in 1977 in translation by Anatol Rapoport. In his recent book Algebra of Conscience, D.Reidel, 1982, Lefebvre describes a mathematical model of ethical cognition which establishes some formal connections between an individual's inner world and his behavior.

Since 1974 Dr. Lefebvre has lived in the USA, and currently he is research psychologist at the School of Social Sciences, University of California, Irvine.

Address: Dr.Vladimir A Lefebvre
         University of California, Irvine
         School of Social Sciences
         Irvine, CA 92717
         USA

DENNIS MEADOWS received his B.A. in chemistry from Carleton College and his Ph.D. in management from MIT, where he served on the faculty for three years. Since 1972 he has been on the faculty of Dartmouth College, Hanover, New Hampshire.

He directs the Resource Policy Center, an interdisciplinary institute that specializes in the application of microcomputers in the design of tools for management education and decision support.

Prof. Meadows has served as a lecturer and consultant on computer simulation and policy analysis in over 20 countries; in the US he has worked for many industrial, federal, and state organizations. He has written five books on aspects of computer-based analysis; he was director of the million dollar development program that led to the creation of the long-range energy forecasting model still used by the US Department of Energy to project energy prices through the end of the century.

Address: Prof.Dennis Meadows
         Resource Policy Center
         Thayer School of Engineering
         Dartmouth College
         Hanover, NH 03755
         USA

HELGA NOWOTNY holds a Ph.D. in sociology from Columbia University, New York. She has been head of the Department of Sociology at the Institute for Advanced Studies, Vienna, and spent one year of research at King's College, Cambridge, before becoming the Executive Director of the UN-affiliated European Centre for Social Welfare in 1974, where she still is at present. She has been guest professor at the University of Bielefeld and in 1981/82 Fellow of the Wissenschaftskolleg zu Berlin. Since 1982 she has also been assistant professor at the University of

Vienna and in 1985 she was elected Chairperson of the Standing Committee for the Social Sciences of the European Science Foundation.

Her major publications are "Kernenergie - Gefahr oder Notwendigkeit", "Nineteeneightyfour: Science between Utopia and Dystopia" (edited with E.Mendelsohn), "Social Concerns for the 80'ies", and numerous articles in social studies of science and science policy.

Address: Dr.Helga Nowotny
European Centre for Social
Welfare Training and Research
Sponsored by the United Nations
Berggasse 17
1090 Vienna
Austria

ROBERT ROSEN: Date of Birth, June 27, 1934; Place of Birth, Brooklyn, New York, USA. Degrees: B.A. (Mathematics), Brooklyn College, 1955; M.A. (Mathematics) Columbia University 1956; Ph.D. University of Chicago (Mathematical Biology) 1959, (N. Rashevsky, Major Professor).

Positions: Research Associate, Committee on Mathematical Biology, University of Chicago, 1960-63; Assistant Professor, 1963-66. Associate Professor, Depts. of Mathematics and Biophysical Sciences, State University of New York at Buffalo, 1966-70; Professor, 1970-76. Associate Director, Center for Theoretical Biology, SUNYAB, 1972-74; Acting Director, 1974-76. Killam Professor, Dalhousie University, 1975-80; Professor of Physiology & Biophysics 1980 - present.

Major Research Interests: Regulation and Control in Biological Systems; Physical Basis of Morphogenesis; Elaboration of Homologies between Biological and Social Systems.

Address: Dr.Robert Rosen
Dept. of Physiology and Biophysics
Faculty of Medicine
Dalhousie University
Sir Charles Tupper Medical Bld.
Halifax, Nova Scotia B3H 4H7
Canada

ROBERT TRAPPL is professor and head of the Department of Medical Cybernetics and Artificial Intelligence, University of Vienna. He holds a PhD in psychology (minor: astronomy), a diploma in sociology, and is engineer for electrical engineering. Since 1970, he has been president of the Austrian Society for Cybernetic Studies, and has been annually reelected in this capacity. In 1984 he was elected president of the International Federation for Systems Research (IFSR). When, in 1984, the Austrian Research Institute for Artificial Intelligence was founded, he was appointed its first director.

The few moments that he is not engaged in "directoral" or "presidential" activities, he writes scientific papers (some 80), edits or co-edits books (14), edits journals ("Cybernetics and Systems", "Applied Artificial Intelligence: An International Journal"), does pantomime, and enjoys life.

Address: Prof.Robert Trappl
        Department of Medical Cybernetics
        and Artificial Intelligence
        University of Vienna
        Freyung 6
        1010 Vienna
        Austria

LEN TRONCALE is Director of the Institute for Advanced Systems Studies at California State University, Pomona. His doctorate was earned in cell and molecular biology. He has taught courses in these specialties as well as evolution, genetics, and systems science. Dr. Troncale is currently Vice-President and Managing Director of the Society for General Systems Research (SGSR), and a member of the Board of Directors of the International Federation for Systems Research (IFSR). He has delivered guest lectures at Universities and Institutes in a dozen countries.

His contribution to this volume was prepared while he was visiting professor at the Department of Medical Cybernetics and Artificial Intelligence at the University of Vienna, Austria.

Dr. Troncale's research interests are in Linkage Propositions between Systems Concepts, Empirical Refinement of Hierarchy Theory, Research on a Theory of Emergence, and Methods of Education in the Systems Sciences.

Address: Prof.Len Troncale
        Institute for Advanced
        Systems Studies
        Calif. State Polytechnic Univ.
        3801 West Temple Avenue
        Pomona, CA 91768
        USA

STUART A. UMPLEBY is an associate professor of management science at George Washington University in Washington, D.C. He studied at the University of Illinois at Urbana-Champaign and received his Ph.D. from there in Communications in 1975 with the dissertation on "Some Applications of Cybernetics to Social Systems."

Between 1977 and 1980 he was principal investigator on a National Science Foundation grant that enabled about 60 cyberneticians and systems theorists from the US, Canada, and Europe to communicate with each other via an electronic information exchange system.

Prof. Umpleby's chief interests are cybernetics and group decision-making methods. He has published widely on these subjects. In addition he has taught courses in system dynamics, artificial intelligence and space industrialization. He is a member of various professional societies and was President of the American Society for Cybernetics from 1980 to 1982.

Address: Prof.Stuart A. Umpleby
        George Washington University
        Dept. of Management Science
        Washington, DC 20052
        USA

FRANCISCO J.VARELA was born in Chile in 1946, and has held a doctoral degree in biological sciences from Harvard University since 1970. His interests have centered on biological and cybernetic mechanisms of cognitive phenomena, especially perception, and the epistemological questions which their study raises. He has contributed more than fifty articles to professional journals on these matters, as well as four books, including "Principles of Biological Autonomy" (North Holland, New York, 1979). Dr. Varela is Professor of Biology at the University of Chile, Santiago, and Senior Researcher at the Center for Epistemological Studies, Ecole Polytechnique, Paris.

Address: Dr.Francisco J. Varela
         Ecole Polytechnique
         CREA
         1, Rue Descartes
         Paris 5
         France

NAME INDEX

Ackoff R.L., 11, 46, 49, 67, 76, 135, 136, 138
Adler A., 15
Airaksinen T., 33
Alexander the Great, 43
Andrews J.G., 7
Aristotle, 9, 39, 43, 67
Ashby W.R., 4, 7, 8, 10, 11, 13, 35-42, 46, 76, 133, 137, 138
Bacon F., 21, 22, 27, 43
Banathy B.H., 102
Baranov P.V., 129
Barr A., 97, 102
Beer S., 46, 76, 121, 122, 133, 139
Beer S.A., 149
Berlinski D.D., 46, 76
Bjorn-Andersen N., 23, 27
Bloch E., 27
Boden M., 25, 27
Boffey P.M., 46, 76
Bohr N., 51
Boulding K.E., 46, 76
Boyle R., 4
Brachman R., 99, 102, 103
Bronowski J., 44, 76
Brown G.S., 125, 130
Buchanan B., 100, 103
Buddha, 17
Bunge M., 47, 53, 76
Canetti E., 5
Capra F., 58, 76
Carter J., 93
Ceccato S., 149
Charniak E., 97, 103
Checkland P., 46, 76
Chestnut H., 102
Chung-Yuan C., 58, 76
Churchman C.W., 46, 76
Ciccarelli E., 102
Cohen P.R., 97, 103
Conant R.C., 7, 8, 10, 11, 13
da Vinci L., 43
Darwin C., 52, 144, 145
Davis J.C., 22, 26, 27
de Chardin T., 13
de la Mettrie J.O., 29
de Mey M., 25, 28
Descartes R., 9, 35, 117, 120
Deutsch K., 46, 76

Druzhinin V.V., 125, 129
Earl M., 27
Easton D., 47, 76
Eigen M., 46, 76
Elzinga A., 20, 27
Enns R.H., 46, 76
Etzioni A., 134, 138
Euler L., 67
Feigenbaum E.A., 97, 102, 103
Findler N.V., 103
Fleck J., 20, 22, 23, 25, 28
Forrester J., 81
Francois C., 29
Gerard, 58, 69
Gesyps R.G., 49, 76
Ghosal A., 31
Greenfeld N., 102
Hai A., 52, 77
Haken H., 46, 77
Hayes-Roth F., 103
Heims S., 117, 122
Heisenberg W., 14
Hinterleitner G., 102
Holst O., 27
Hoos I.R., 46, 77
Horn W., 49, 78, 102, 103
Hu N.C., 33
Hume D., 14, 111
Iberall A.S., 46, 77
Jain V., 49, 50, 53, 77
Jamison A., 20, 27
Jantsch E., 47, 77
Jesus, 17
Jung C.G., 5, 13
Kant I., 113, 114, 115, 140
Kennedy J.F., 134
Kepler J., 32, 45
Khrushchev N., 134
Klir G.J., 46, 49, 52, 76, 77, 78
Kobsa A., 99, 103
Koestler A., 47, 77
Koselleck R., 21, 28
Kraft R.W., 101, 103
Kruskal M., 46, 80
Kuratowski K., 67
Lefebvre V., 123, 124, 126, 129, 130, 139, 150
Leibnitz W., 67
Leinfellner E., 103
Lenat D.B., 103
Lepsky V.E., 130
Levesque H., 102
Lincoln A., 88
Locke J., 9
Lucretius, 58, 77
Luhmann N., 21, 28
Majone G., 46, 77
Mandelbrot B.B., 46, 77
Manuel F.E., 20, 28
Manuel F.P., 20, 28
Marbach E., 113
Marker C., 19

Maturana H.R., 6, 46, 77, 129, 130
McCulloch W., 9, 109, 116, 141
McDermott D., 97, 103
Meadows D., 81
Meadows D.L., 150
Mendelsohn E., 20, 28
Mersenne M., 108
Merton T., 58, 77
Miller H., 66, 77
Miller J.G., 46, 49, 50, 52, 53, 67, 77
Mohammed, 17
Montaigne M., 108
More T., 44
Mumford E., 27
Negrotti M., 23, 28
Neurath O., 20, 28
Newton I., 14, 35, 37, 38, 39, 40, 41, 42, 146
Nilsson N.J., 102
Nowotny H., 20, 21, 28-34, 150
Ockhams W., 50
Ohm G.S., 4
Oliva T.A., 50, 53, 78
Oren T.I., 52, 77
Parsons T., 46, 78
Pask G., 67, 78, 108, 116
Piaget J.R., 108, 110, 115, 116
Pierce C., 67, 141
Pigeman V., 102
Plato, 109
Popper K., 59, 78, 143
Powers W.T., 108, 116, 121, 122
Prendergast K.A., 103
Prigogine I., 46, 78
Quade E.S., 46, 77
Rapoport A., 47, 78, 150
Robbins S., 50, 53, 78
Rogers G., 49, 76
Rosen R., 46, 78, 151
Russel B., 38
Sadovsky V., 99, 142, 143
Schaffner K.F., 50, 67, 78
Schank R.C., 99, 103
Schatz W., 98, 103
Schmid K., 102
Schmolze J.G., 99, 103
Schreider Y.A., 130
Schuster P., 46, 76
Shortliffe E.H., 100, 103
Simon H.A., 46, 78
Smith E.T., 103
Smythies J.R., 47, 77
Socrates, 39, 107
Steinacker I., 99, 103
Thom R., 46, 78
Toom A.L., 130
Toynbee A., 54
Trappl R., 10, 20, 28, 49, 78, 97, 99, 103, 133, 151
Troncale L.R., 33, 34, 43-46, 50-79, 102, 152
Trost H., 99, 102, 103
Trudoliubov A.F., 129, 130, 131
Turing A., 117, 118, 119, 120, 121, 146, 147

157

Turkle S., 25, 28
Umpleby S.A., 49, 99, 135, 138, 139, 140, 147, 152
Varela F., 6, 46, 77, 118, 120-125, 129-131, 143-147, 153
Venn J., 67
Verity J.W., 98, 103
Vico G., 108
von Bertalanffy L., 46, 76
von Foerster H., 46, 49, 76, 124, 125, 131
von Glasersfeld E., 110, 112, 116, 139-144, 149
von Goethe J.W., 32
von Linne K., 52
von Neumann J., 9, 67, 117, 118, 120, 122
Voorhees B., 43, 46, 67, 68, 79
Waddington C.H., 35, 46, 79
Waldrop M.M., 24, 28
Warfield J.N., 46, 49, 67, 79
Waterman D.A., 98, 103
Weber M., 145
Weizenbaum J., 23
Whyte L.L., 46, 80
Wiener N., 46, 67, 80, 117, 120, 122
Wilber K., 58, 80
Wilson A., 46, 61, 67, 80
Wilson A.G., 50, 59, 79
Wilson D., 46, 80
Winner L., 24, 28
Winston P.H., 97, 103
Winter M., 21, 28
Yonke M., 102
Young O.R., 80
Young T.Y., 49
Zabusky N., 46, 80
Zeleny M., 129, 131
Zeller C., 102
Zwicky F., 61, 67, 80

SUBJECT INDEX

Abstraction mapping utility, 71
ACE, 101
Adaptation, Ashby's theory of, 133
Adaptiveness, 35
Advisory council, 92
Alcohol addiction, 94
Alcoholics Anonymous, 15
Algebra of conscience, 128, 150
Ambiguity, 110
Analysis, strategic, lifecycle of, 87
Antibody, 119
  anti-idiotypic, 120
Anticipation, 39, 41
Applied Artificial Intelligence: An International Journal, 151
Approach, mechanistic, 35
Army, 123
Artificial Intelligence, 22, 23, 24, 27, 29, 31, 97
  curricula, 97
  definition, 97
  groups, inhouse, 98
  ontology, 99
  tools, 102
Aspiring individual, 8
Association classes of Linkage Propositions, 57
Atomist, 39
Austrian Research Institute for Artificial Intelligence, 151
Austrian Society for Cybernetic Studies, 151
Automation, 16
Autonomy, 43
  cybernetics of, 117
  definition of, 117
Autopoiesis, 6, 129
Awareness, 126
Background noise, 118
Behavior, 36
  adaptive, 133
  telic, 37
Bifurcations, 40
Biology, 36
Biosphere, 13
Bottom-up, 61
Brain, 9
California, 15
California Institute for Advanced Systems Studies, 152
Canada, 16
Capable self, 8
Carnegie-Mellon University, 101

Cartesian causality, 117
Categorical Imperative, 114
Categories of causation, 39
Category switching, 69
Causal order, 38
Causality, 38
Causation
  Aristotelian categories of, 39
  material, 39
Cause, 38
  efficient, 39
  final, 39
  formal, 39
Caveat about standards, 95
Center for Cybernetics, 149
Center for Epistemological Studies, Ecole Polytechnique, Paris, 153
Change
  causal model of, 14
  nature of, 14
Co-operative properties, 118
Communication, tool for, 84
Community, 10, 11
Complexity, physics of, 35
Computer-based analysis, 150
Computerized data bases, 64
Conceptual fit, 111
Conflicting structures, 123, 126, 150
Connectedness, 69
Consciousness, 6, 24
Constitutive parameter, 39
Construction of reality, 111
Constructivism, 113
Constructivist epistemology, 114
Constructivist theory of knowledge, 114, 149
Contemporary physics, 40
Context, 68
Control, 31, 108, 117
  centralized, 23
Control ideology, 122
Control of perception, 121
Corporate clients, 135
Correspondence principle, 67
Corroboration by others, 111
Country, 11
Criminals, 15
Crisis Handling Expert System, 100
Critique of Pure Reason, 113
Cross-level hypotheses, 49, 50, 67
Cuban missile crisis, 134
Cybernetic Society, 10
Cybernetician, 12
  American, 99
  Soviet, 99
Cybernetics, 4, 123
  first order, 125
  of the Soviet Union, 127
  second order, 123, 125, 128, 129
  Western, 127
Cybernetics and Systems: An International Journal, 151
Cybernetics of autonomy, 117
Dalhousie University, 151

Dartmouth College, 150
Deabstraction, rules for, 62
Decision maker, 89, 137
Decision making process, strategic, 85
Defense Advanced Research Projects Agency (DARPA), 98
Definition of isomorphies, 58
Dehumanizing effect, 23
DENDRAL, 101
Department of Medical Cybernetics and Artificial Intelligence, University
        of Vienna, Austria, 151
Deterrence, 15
Developing nation, 13
DIPMETER ADVISOR, 101
Discinym, 56
Dissemination of results, 94
Dynamic process-orientation, 59
Dystopian vision, 22
Earth, 13
Eastern philosophy, 14
Economics, conventional, 52
Eigen-behavior, 118
Electronic brain, 9
Emergence of meaning, 118
Empirical refinement, 51
Empty formalism, trap of, 68
English-Russian/Russian-English translation program, 99
Entailment network, 67
Entelechy, 9, 11, 17
Entelechy for the nation, 16
Entitation, 58, 69
Epistemology, 37, 108, 109, 110, 137
  dichotomous, 25
  mechanistic, 37
Equilibrium-thinking, 32
Equivalence, 110
Ethics, 113, 114
European Centre for Social Welfare, 150
European Strategic Programme for Research and Development in Information
        Technologies (ESPRIT), 98
Evolution, 52
Experiential compatibility, 111
Experiential reality, 114
Experiential world, 112
Expert system, 23, 97, 101, 103
  military, 98
Explanation part, 101
Extended cognitive space, 25
Facetism, 69
Feedback loop, 133, 137
Final causation, 41
Forest damage, 83, 84, 85, 87, 89, 90, 91, 93
FORTH, 71
Fraud, 87
Funding, 91
Future collapse, 136
Game theory, 17
Gea-hypothesis, 13
General systems literature, bibliography of, 49
General systems problem solver, 52
General systems science, 45, 59
General systems science, reconceptualization of, 47

George Washington University, 99, 152
Geosphere, 13
Goal-directed action, 108
Graduated entry, 63
Group decision-making methods, 152
Harmony, 21, 32
Health, 10
Heuristic learning program, 64
Hierachy theory, 152
Holism, 57
Holistic intellectual movement, 52
Hologram, 10, 14
Homeostasis, 13
Hospital, 10, 12
Hot-line, 100
Human systems engineering, 43
Idealism, 115
Identification of issue, 90
Idiotype, 120
Image, formal, 38
Imaging process, 38
Immune system, 120
Immunology, 119
Implementation, 90
Implementation process, conventional view, 89
Implication, 38
In-formation, 122
Inconsistencies, 84
India, 12
Individual identity, 110
Individual, model of the, 7
Inference engine, 71, 100, 101
Influence vector, 56
Information, 122
Initial condition, 39
Inner circle, 89
Institute for Advanced Studies, Vienna, Austria, 150
Institute of Cultural Affairs, 135, 136
Institution, 10, 11
Institutional setting, 91
Institutionalized revolution, 12
Instructive mode of interaction, 119
Instrumentalist, 115
Insurrection, 17
Integral operator, 39
Intercultural knowledge base, 99
International Federation for Systems Research (IFSR), 49, 151, 152
International Institute for Applied Systems Analysis (IIASA), 89
International Monetary Fund, 16
International Society for Ecological Modelling, 61
International tension, 97, 98
Invariance, 4
Invisible college, 93
Isomorphies, 43, 46, 48
  anasynthetic, 51
  enhanced search for, 59
  taxonomy of, 52
Japanese society, 19, 24
Kaypro computer, 66
Kennedy Experiment, 138
KL-ONE, 99

Knowledge, 107, 115
  concept of, 108
  special, 44, 46
Knowledge acquisition component, 101
Knowledge base, 71, 101
Knowledge representation, 97
KRYPTON, 99
Language, 111
  Newtonian, 37
Language understanding, 97
Law, 11
Law of requisite variety, 7
Laws of form, 125, 126
Laws, spatio-temporal, 14
Leadership Effectiveness and New Strategies (LENS), 135
Learning, 97
Legitimation of a team, 92
Leisure time, 32
Level of reality, 111
Lifework integrator utility, 66
Linkage Proposition, 43, 152
  definition of, 55
  empirical refinement of, 59
Linkage Proposition Template Model (LPTM), 43
  graphic, 64
  mathematical formalization of, 67
LPTM and graph theory, 69
LPTM as an expert system, 70
LPTM Rule-Based System (GENSYS), 71
LISP-machines, 101
Love, falling in, 31
Macintosh
  computer, 63
  graphics, 66
Management cybernetics, 10
Management training games, 89
Manchester University, 149
Mandala, 5, 14, 57
Mathematics, 38
Mechanics, Newtonian, 37
Mechanism of variation, 34
Mechanisms, 41
META-GENSYS, 71
Meta-rules, 100
Meta-taxonomy, 69
Metapraxis Ltd., 149
Mexico, 12, 15
Micro-utopia, 20, 31
Microchips, 17
Mid-range missile, 102
Military
  applications, 97
  simulation, 98
Mind, 25
Mode of inference, 119
Model builder, 84
Model of coal production, 88
Model of yourself, 8
Models
  ecological-economic, documentation of, 85
  strategic, 82, 83, 87

    strategic, criteria of, 87
    strategic, implementation of, 85
    tactical, 82
    unfinished, 87
Molecular biology, 40
Molecular genetics, 101
MOLGEN, 101
Moscow State University, 150
Multi-person game, 89
MYCIN, 100
Nation, 11
Nation, entelechy of, 13
Natural language systems, 98, 99
Natural Law, 37, 38
Nervous system, autonomic, 7
Nervous system, central, 7
Neurocybernetics, 9
New technology, adoption of, 134
Nineteeneightyfour, 151
Noosphere, 13
Nuclear Test Ban Treaty, 134
Object permanence, 110
Objectivity, 107, 112, 113, 115
Observable, 36
Office for Emergency Preparedness, 88
Oil depletion, 90
Om-word, 57
Ontological reality, 115
Ontology, 99, 109
Open system, 40
Operational closure, 118
Oppression, 5
Origins of isomorphies, 58
Ostensible self, 8, 9
Pantomime, 151
Paradigmatic position, 37
Particle physics, 51
Pennsylvania University, 149
Penology, 15
Phenomenon of man, 13
Physical appearance of a report, 92
Physics of complexity, 35
Pilot studies, 102
Planetary level, 13
Planning, 97
Politics, 84
Potential self, 9
Power, 5, 43
  recursions of, 3
Predictions, 38
Press relations, 93
Principles of biological autonomy, 153
Progress, belief in, 21
Proposition, 38
Purpose of systems, 12
Railway, 10
Real world knowledge base, 99
Realist, 110
Reality, 107, 108, 126
  construction of, 110
Reasoning, 97

Reconstructability Theory, 52
Recruiting partners, 93
Recursions, 6
Reductionism, 4, 38
Reductionist experimentation, 61
Reductionistic position, 37
Reflexion, 128
Reflexive analysis, 123, 126, 127
Reflexive control, 123
Representation, 119
Requisite variety, 12
Research management tool, 61
Resource Policy Center (RPC), 81, 89, 92, 95
Restructuring organizations, 135
Robot, 23, 98
Robust interaction, 51
Rock of Gibraltar, 4
Rule-base, 71
Rules, 100
Sans Soleil, 19
Satori, 14
Scale translation protocols, 62
Schedule for a work, 94
Science, 19
Science of utopistics, 20
Second order cybernetics, 123, 137
Self, idea of, 5
Self-awareness, 6
Self-definition, 51
Self-determination, 118
Self-objectification, 123
Self-organization, 118
Self-production, 6
Self-reference, 6, 125, 139
Self-reflexion, 125
Self-regulation, 16, 107, 108
Self-reproduction, 6
Self-survival of parties, 12
Sensitivity analysis, 84
Simulation models, 81
Skin, 7
Social engineer, 20
Social order, 22
Social organizations, 133
Social policy, 81
Social ritual, 19, 24, 27
Social science, 134
Social system, 5, 10
Social utopia, 22
Society for General Systems Research (SGSR), 58, 71, 152
Software mistakes, 86
Solipsism, 115
Soviet Union, 123, 125
Soviet-American relations, 134
State, 36
Stepwise refinement, 63
Strategic computer, 81
Strategic Computing Program, 98
Strategic discussion, 88
Structure of awareness, 126, 150
Survival, 5

Symbolic studies of relations, 67
Synergetics, 118
Synergy, 4
Synthesis, 4
Synthesis in science, 4
System
  biological, 117
  bootstrapping, 63
  complex, 41
  definition by Ashby, 36
  dynamical, 38, 39, 41
  human, 45
  man-made, 43
  material, 35
  mathematical, 38
  natural, 35, 43
  simple, 41
  social, 134, 135, 137
  stable, 44
  state-determined, 36, 37, 38, 41
    simplest, 36
System dynamics, 81
System of systems concepts, 49
System properties, continuity of, 40
Systemic failure, 12
Systemness, 44
Systems
  complex, science of, 41
  input/output, 122
  invariant properties of, 4
  life cycle of, 54
  rule-based, 70
  simple, science of, 41
Systems autonomy, 44
Systems concept, 49
Systems environment, 69
Systems resonance, 69
Systems science, 4, 43, 45, 46, 53
Systems terms, glossary of, 49
Systems theory, 117
Tactical decision making, 98
Taxonomy
  human-usage-based, 53
  phenomenologically-based, 53
Taxonomy of isomorphies, 45, 52
Teaching tool, 61
Team-shared research outlines, 66
Technological optimism, 21
Technology, 20
Technology assessment, 20
TELENET, 100
Teleology, 37
Telepathy, 4
Telephone networks, 101
Telos, 39
Tension, international, 97
Theorem of Conant-Ashby, 7, 11, 13
Theoretical physics, 40
Theoretical science, 38
Theory of emergence, 152
Theory of knowledge, 107

Thinking machine, 23
Third World, 13
Threat assessment system, 101
Threshold summation, 69
Time, conceptualization of, 21
Toolbox, 62
Top-down, 61
Torture, 5
Transdisciplinary research, 66
Travel system, 10
Truth, notion of, 109
Turing machine, 117
TYMNET, 100
Typing mistakes, 86
Understanding, mutual, 97
Unemployment, 16
Union of International Associations, 49
Universe, reductionist model of, 4
University, 11
  of California, Irvine, 150
  of Chile, Santiago, 153
  of Georgia, 149
  of Vienna, Austria, 151
Universum, 123, 125
USA, 98, 99, 100
U.S. Central Command, 98
U.S. Navy, 101
User interface, 101
USSR, 99, 100
  Academy of Sciences, 99
Utilitarianism, 115
Utopia, 13, 19, 44
Utopian-dystopian dichotomy, 23
Utopian-dystopian tension, 27
Variability, 34
Variable, 36, 40
Venture capital, 98
Viability, 107, 109, 114
Vision, 97
War, 101
War machines, 43
Work time, 32
World Organization of General Systems and Cybernetics (WOGSC), 149
World-consciousness, 13
XCON, 101
Zen-Buddhism, 14